300 ANIMALS IN FULL COLOR

POISONOUS

A

a G

Ori

ED. **J. BRODIE, JR., Ph.D.**

Illustrated by
JOHN D. DAWSON

St. Martin's Press ♊ New York

FOREWORD

Just as previous Golden Guides have served to introduce young people (including the author of this guide) to science, it is hoped that this guide will stimulate interest in the spectacular evolution of venomous animals and their use of venoms to subdue prey and repel predators.

The artist, John D. Dawson, with the able assistance of his wife, Kathleen Dawson, has created the finest collection of venomous animals illustrations ever produced. Thanks are also due Caroline Greenberg, Senior Editor, for her help in developing the entire concept of this guide and for her attention to detail.

I am indebted to a number of colleagues for their advice and for materials from which to illustrate this book, especially: Jonathan A. Campbell; Daniel R. Formanowicz, Jr.; William Lamar; Robert F. McMahon; Jim Stout; Edmund D. Brodie, III; Ronald A. Nussbaum; Jay Vannini; George Fichter; Remo Cosentino; and David Barker. John P. Nelson prepared the scanning electron micrographs at the University of Texas at Arlington Center for Electron Microscopy. I am grateful to Judith Johnson Brodie for her continuous support.

E.D.B.

CONTENTS

VENOMOUS ANIMALS

People are strangely attracted to dangerous, potentially deadly animals. At a zoo, watch how many people congregate at the venomous snake exhibit. The conflicting reactions of fear and fascination are clearly evident.

This book is an introduction to the marvelous diversity and beauty of venomous animals, concentrating on spectacular and deadly species. Although only a few of the tens of thousands of venomous animals can be described and illustrated here, all major groups in the world are represented, including the most dangerous in the animal kingdom. The variety of uses and origins of venoms are indicated. Not included are those animals that are toxic when eaten, such as the hundreds of species of fishes and shellfishes—unless, as in the case of some amphibians and insects, would-be predators are repulsed by their toxins.

VENOM In general terms, venom refers to a substance used by one animal to cause injury or death to another. More narrowly, venom refers to substances delivered either by biting or by stinging. These animals are referred to as *actively venomous*. Actively venomous animals have a *venom apparatus*—a means for dispensing their venom into other animals. Some inject the venom through hollow, hypodermiclike

Stone
Fish
spine

Spiney Newt spine

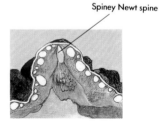

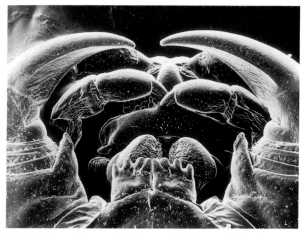

Centipede's fangs, photographed with a scanning electron microscope

teeth or stingers. Others give off venom from spines that puncture or rip into a victim.

In contrast, *passively venomous* animals are those with secretions that affect other animals when eaten. These animals are usually referred to as *poisonous*. Passively venomous animals cannot inject venom into other animals. Some of the insects and amphibians are passively venomous.

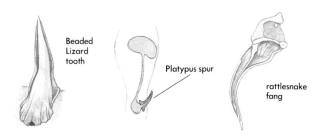

Beaded Lizard tooth

Platypus spur

rattlesnake fang

5

HOW VENOMS SERVE ANIMALS

SUBDUING PREY Using venom to paralyze or kill enables animals to attack and eat prey much larger than they otherwise could without risking injury. In many of these animals the venom gland is a modified salivary gland with a duct leading to a hollow or grooved venom apparatus, commonly a specialized or enlarged tooth or fang. Venomous snakes, shrews, octopuses, centipedes, snails, spiders, and some insects are of this type.

Cnidarians have specialized stinging cells in their tentacles, while scorpions possess a unique venom apparatus at the tip of their tail. Some wasps use their stinger (an ovipositor found only in females) to paralyze prey that is to be eaten by the wasp's larvae.

REPULSING PREDATORS Animals that employ venom to subdue prey may also use it as a defensive mechanism against their own would-be predators. Other animals use venom to defend themselves (either actively or passively) but do not use the venom to subdue prey. These antipredator adaptations are most effective against predators that learn to avoid the repulsive prey. Sea urchins, various fishes, a salamander, the Platypus, and some insects (such as bees, wasps, and caterpillars) are among the actively venomous animals using venom only to repulse predators. Spitting cobras, a variety of insects, a salamander, and others spray venom in order to repulse predators at a distance. Certain amphibians and some fishes produce toxic secretions in their skin glands. Others, primarily insects, concentrate poisons from plants in their bodies and in this way render themselves inedible. A few animals, such as nudibranchs and hedgehogs, utilize venom of other animals in their own defense.

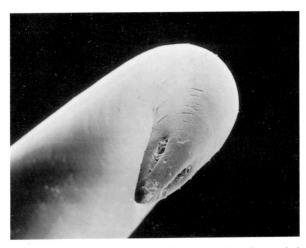

Tip of the stinger of a scorpion showing openings of venom ducts, photographed with a scanning electron microscope

Hedgehog rubbing a venomous toad onto its own spines — making *them* venomous

DANGER FROM VENOMOUS ANIMALS

As highly dangerous venomous animals are most abundant in less industrialized nations, their bites and stings are often not reported. Even deaths sometimes remain unrecorded.

The most reasonable estimate of venomous snakebite deaths worldwide is 40,000 to 50,000 per year, but this information is neither complete nor accurate.

In India, for example, an estimated 200,000 people are bitten by venomous snakes each year. Of these, 10,000 to 15,000 die. In Burma, although we know that about 15 of every 100,000 people die each year of venomous snakebites, we don't know how many are *bitten*.

SNAKES THAT CAUSE THE MOST DEATHS IN THE WORLD

Malayan Pit Viper
Southeast Asia

Common Lancehead
South America

Puff Adder
Africa

Better records are kept in the United States, but the total number of bites is almost certainly not reported. Records showing fewer than 7,000 bites and 15 deaths from venomous snakes per year indicate a mortality rate of less than 1 per 10 million. These figures show that snakebites pose little threat to life in the United States. Most deaths are from bites of diamondback rattlesnakes.

Bees and wasps account for approximately 25 deaths every year in the United States, thus posing a greater threat than do snakes. Spiders and scorpions are responsible for another five deaths each year. It is impossible even to give a reasonable estimate of the number of deaths caused worldwide by venomous creatures other than snakes.

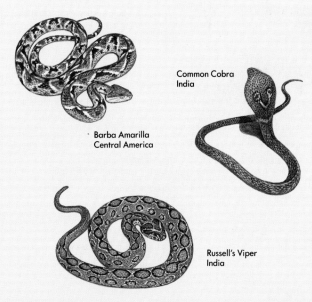

Common Cobra
India

· Barba Amarilla
Central America

Russell's Viper
India

THE ACTION OF VENOMS

Some venoms, especially those from passively venomous animals, burn the tender mouth lining and eyes of predators (sometimes resulting in temporary blindness). Other venoms cause pain or illness to would-be predators.

Actively venomous animals, however, possess the most dangerous venoms. These can be generally categorized as:

    1) *neurotoxins*, which affect the nervous system and cause death by paralysis

    2) *hemotoxins*, which digest tissues—including the blood cells

    3) *cardiotoxins*, which affect the heart directly

Venoms are actually far too complex to be divided strictly into these categories, however. In fact, many animals have venoms with components of two or of all three of these actions.

FIRST AID A person bitten or stung by a venomous animal should seek medical care as soon as possible. Because it is important for the doctor to know the animal responsible for the bite or sting, it may be necessary to kill the animal and take it to the doctor with the victim. Keep in mind that animals believed to be dead may still be capable of mustering a defensive bite or sting, so be extremely careful.

If a hospital cannot be reached within 30 minutes or if a child is bitten by a large venomous snake, it may be necessary to apply a constriction band above the bite to prevent or slow the spread of the venom. Care must be taken not to cut off the flow of blood.

Only a person with medical training should attempt to cut and remove venom from a bite. The "cut and suck" method of treating snakebite, when improperly done, can cause more harm than the venom.

Western Newt
defensive posture

Sandhills Hornet

WHY VENOMOUS ANIMALS
ARE BRIGHTLY COLORED

Venomous animals are commonly more conspicuous than nonvenomous ones. Red, orange, yellow, or metallic colors, or any of these colors in banded or spotted patterns with black and white make them highly visible. It is advantageous for nonvenomous animals to be inconspicuous (cryptic) in coloration, thereby avoiding detection by predators. But venomous species advertise their presence to other animals. Most beneficial is a color pattern that is concealing to an animal's prey but a warning (aposematic) to its predators. This is true, for example, when a venomous animal preys upon animals without color vision, but is a potential prey of animals with color vision.

Many venomous amphibians are cryptically colored on the back, but when attacked by a predator, turn their body to expose a bright coloration on their belly and on the soles of their feet. Predators soon learn to avoid these warning colors following painful contact with a venomous animal. Predators that have learned to avoid a brightly colored venomous animal (the model) will also avoid a similarly colored nonvenomous animal (the mimic). Some mimics of venomous animals are shown on pages 64-65 and 144-147.

USES OF VENOMS

Since ancient times venoms have been used in weapons of warfare and for hunting. Early warriors in Europe as well as some North American Indians coated arrowheads and spears with snake venoms. Roman and Greek warriors reportedly threw containers of venomous snakes into the boats of their enemies. Today some tribes in Africa poison their arrows with insect venom, while South American tribes poison their blowgun darts with venom from poison dart frogs.

RESEARCH The nature of venoms—neurotoxic, hemotoxic, and cardiotoxic—makes them useful in understanding the functioning of the human body. Much of our knowledge about the workings of the nervous system is the result of research with venoms. With the help of genetic engineering technology, venoms are now being examined for possible value in pest control and as antimicrobial agents.

ANTIVENIN Venoms are utilized in making antivenin products. Antivenins are produced by injecting a series of small doses of venom into a large animal, such as a horse. As soon as antibodies have formed in the horse's blood, rendering it resistant or immune to the venom, blood is drawn and the serum removed. This serum is the basis for the antivenin's effectiveness against the effects of a venomous bite.

Antivenins are available for Stone Fish, several spiders (including Black Widows and funnel web spiders), some scorpions, and a wide variety of snakes. In the United States, a single antivenin is effective against the bites of rattlesnakes, Cottonmouths, and Copperheads. Other dangerous snakes generally require specific antivenins because their venoms are so different that the antibodies formed against one are not effective against another venom.

Some South American Indians coat their darts with frog venoms

MEDICAL venoms, because they have specific effects on the human body, have been used in medical treatment throughout history. Both the ancient Egyptians and the early Chinese utilized venoms in treating numerous medical disorders. Early Chinese medical books listed hundreds of uses for snake venoms and snakes.

Today hemotoxins, usually from vipers, are employed as anticoagulants, while neurotoxins, generally from cobras, are used for the treatment of pain. Snake venoms are also being used on an experimental basis in treating diseases of the nervous, cardiovascular, and musculoskeletal systems. In addition, they are employed in working with some kinds of cancer, for a wide range of diagnostic tests, and as antiviral and antibacterial agents.

In China, toad venom is collected, dried, and sold as Ch'an Su, which is used for treating heart disease and a number of other ailments. Some scientists believe that a component of bee venom is a therapeutic agent in the treatment of arthritis, but its effectiveness has not yet been demonstrated.

13

CNIDARIANS

About 9,000 species of cnidarians—corals, jellyfishes, and sea anemones—inhabit the seas of the world. A few species live in fresh waters. All have tentacles armed with thousands to millions of darts, called nematocysts, that are fired into prey animals, the venom paralyzing or killing them.

Most cnidarians are not dangerous to humans, the venom from their nematocysts causing only local pain and burning, but a few—none of the freshwater species—have extremely potent venom that can cause death within a few minutes. People should never pick up jellyfish along beaches. The tentacles sting even after the jellyfish is dead; nematocysts can penetrate rubber gloves and thin clothing. A person who is stung should get out of the water, scrape the tentacles off the skin with a stick, and get medical attention immediately.

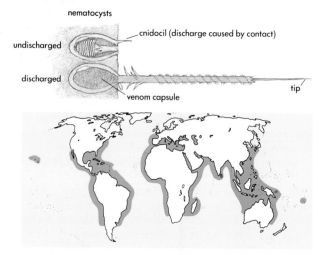

nematocysts

cnidocil (discharge caused by contact)

undischarged

discharged

venom capsule

tip

Areas where dangerous cnidarian stings are most likely

HYDROIDS, like the Portuguese Man-o-War and the stinging corals, discharge a venom that causes intense pain, a rash, and nausea—rarely death. The Portuguese Man-o-War is the most dangerous cnidarian along the coasts of North America. It is often eaten by the Loggerhead Turtle, which is apparently not susceptible to the venom.

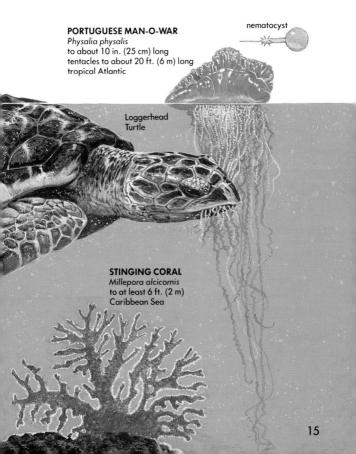

nematocyst

PORTUGUESE MAN-O-WAR
Physalia physalis
to about 10 in. (25 cm) long
tentacles to about 20 ft. (6 m) long
tropical Atlantic

Loggerhead
Turtle

STINGING CORAL
Millepora alcicornis
to at least 6 ft. (2 m)
Caribbean Sea

15

JELLYFISHES are marine pelagic cnidarians that occur in all seas of the world. The 200 species range in size from a few inches to more than 6 feet in diameter. The largest have tentacles as much as 120 feet long. All jellyfishes sting with nematocysts, but in most, the venom is weak and not dangerous to humans. A few species—most of them graceful swimmers that feed on fish—are highly venomous and dangerous. Mild stings cause swelling and blisters, but potent stings cause muscular spasms, respiratory failure, and even death. Most venomous is the Sea Wasp that can kill a human in less than a minute. In Australia, about 9 percent of the stings from the jellyfish are fatal. Some fish are resistant to the venom of jellyfishes and seek shelter among their tentacles.

believed occasionally responsible for human deaths

SEA WASP
Chironex fleckeri
to 3 in. (75 mm) in diameter
Indian and South Pacific
oceans

ANGLED HYDROMEDUSA
Gonionemus vertens
to ⅝ in. (15 mm) in diameter
North Pacific

**LION'S MANE or
SEA BLUBBER**
Cyanea capillata
to 8 ft. (2.4 m) in diameter
occurs worldwide,
as do similar species

SEA ANEMONES live in oceans throughout the world and are often abundant in intertidal areas. They spend most of their time attached to the bottom but can release themselves to creep along the bottom or to float. Many species are brightly colored. With the nematocysts on their tentacles, sea anemones are able to paralyze small animals, but their stings are seldom dangerous to humans. Stings usually cause only itching or burning but occasionally result in swelling or an open, slow-healing sore.

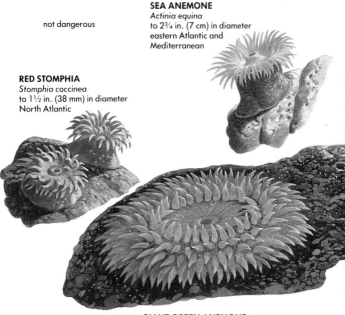

not dangerous

SEA ANEMONE
Actinia equina
to 2¾ in. (7 cm) in diameter
eastern Atlantic and
Mediterranean

RED STOMPHIA
Stomphia coccinea
to 1½ in. (38 mm) in diameter
North Atlantic

GIANT GREEN ANEMONE
Anthopleura xanthogrammica
to 10 in. (25 cm) in diameter
Pacific Coast of Central and North America

MOLLUSCS

Snails, slugs, squid, and octopuses are all molluscs. About 95,000 species of molluscs live in marine, freshwater, and terrestrial environments, but all known venomous species are marine. Several aggressive predatory snails, such as the whelks, have venomous saliva but no specialized venom apparatus. The venom flows from the salivary glands into the wound caused by the rasping mouthpart, or radula.

A number of marine snails are tetrodotoxic (see p. 82), probably from eating toxic microorganisms. Other snails and slugs produce a slime on the skin that repulses potential predators.

CONE SHELLS are the most venomous snails. Most of the roughly 500 species live along coral reefs in the South Pacific. They burrow by day and feed by night, often on small fish that are quickly paralyzed by their venom. These snails have a highly specialized radula—a series of hollow radular teeth (darts) that are filled with venom from the venom gland. These darts are fired one at a time. A fine thread tethers each dart to the snail, which then pulls in its catch like a fisherman. Each tooth is used only once and is immediately replaced by one of several ancillary teeth.

The effect of the venom on people may vary from the strength of a wasp sting to paralysis and death. The toxin affects the muscles directly, and death is caused by heart failure. Reports suggest that as many as 25 percent of stings from some species are fatal, some within minutes. Among the most venomous and having accounted for human deaths are the Geographer Cone and the Textile Cone. The bites of some of the cone shells, such as the California Cone, that feed on worms or on other molluscs are painful but probably not dangerous to humans.

TEXTILE CONE
Conus textile
to about 4 in. (10 cm)
South Pacific

life threatening

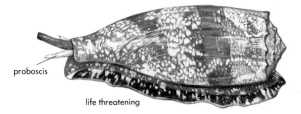

proboscis

life threatening

radular tooth of cone shell

GEOGRAPHER CONE
Conus geographus
to 5 in. (13 cm)
South Pacific

COURT CONE
Conus aulicus
to 6 in. (15 cm)
Indian Ocean

life threatening

CALIFORNIA CONE
Conus californicus
to 1⅝ in. (41 mm)
central California to
Baja California

not dangerous

NUDIBRANCHS, often called sea slugs, are marine snails without shells. Most species eat venomous cnidarians. Some species, like the Sea Lizard, even feed on the Portuguese Man-o-War. They swallow the nematocysts whole, and these unfired nematocysts accumulate in areas of the nudibranch's body surface, usually along the edges or in the often brightly colored projections or fringes. These pilfered nematocysts then serve as defense weapons for the nudibranchs.

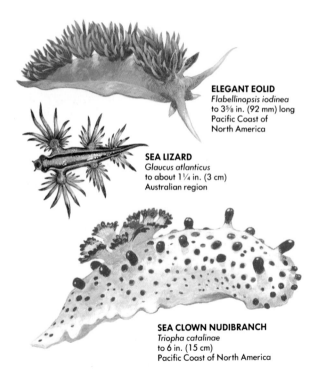

ELEGANT EOLID
Flabellinopsis iodinea
to 3⅝ in. (92 mm) long
Pacific Coast of
North America

SEA LIZARD
Glaucus atlanticus
to about 1¼ in. (3 cm)
Australian region

SEA CLOWN NUDIBRANCH
Triopha catalinae
to 6 in. (15 cm)
Pacific Coast of North America

OCTOPUSES and squids have mouthparts forming a horny beak. Of the some 650 species, only the Blue Ringed Octopus from Australia has dangerously venomous saliva. The venom is the neurotoxin tetrodotoxin (p. 82) that paralyzes prey and can cause numbness, paralysis, and death in humans. Other octopuses may inflict a bite but are not known to be venomous.

BLUE RINGED OCTOPUS
Hapalochlaena maculosa
span of tentacles to about 6 in.
(15 cm)
Australia

beak of octopus

NEMERTEAN WORMS

These marine worms, about 750 species, live in intertidal areas around the world. Also known as ribbon worms, they range in size from a few inches to nearly 100 feet long. Many species are red, yellow, green, white, or combinations of these bright colors. They have a long proboscis that can be everted and extended. In some species the proboscis is armed with a venomous bristlelike stylet used to paralyze prey, often other worms. The venom is apparently produced by mucous glands in the proboscis sheath. Some species produce a neurotoxin; others produce heart toxins. Nemerteans are not known to be dangerous to humans.

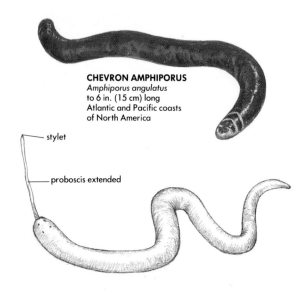

CHEVRON AMPHIPORUS
Amphiporus angulatus
to 6 in. (15 cm) long
Atlantic and Pacific coasts
of North America

stylet

proboscis extended

SEGMENTED OR ANNELID WORMS

An unknown number of species of segmented marine worms are venomous to humans. Most of the about 8,000 marine species are harmless, but at least one—the Blood Worm—has a venomous bite, causing intense pain. A Blood Worm's proboscis is 20 percent the length of its body and has four fangs at its tip. Each fang is connected to a venom gland.

Other worms found worldwide and generally referred to as Bristle Worms or Fire Worms have hollow, venomous spines (setae) on their legs (parapodia). Being jabbed by setae causes burning, swelling, and numbness, but secondary infection is the greatest danger.

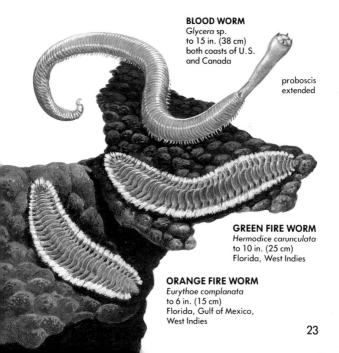

BLOOD WORM
Glycera sp.
to 15 in. (38 cm)
both coasts of U.S.
and Canada

proboscis
extended

GREEN FIRE WORM
Hermodice carunculata
to 10 in. (25 cm)
Florida, West Indies

ORANGE FIRE WORM
Eurythoe complanata
to 6 in. (15 cm)
Florida, Gulf of Mexico,
West Indies

23

ECHINODERMS

Echinoderms, or spiny-skinned animals, are a group of about 6,000 species, all marine and including the star fishes, sea urchins, sand dollars, and sea cucumbers. They move about on large numbers of tubefeet along grooves on their body's arms or divisions, and they repel predators by toxins produced in the glandular skin covering their exoskeleton.

Some of the roughly 2,000 species of starfishes produce toxins painful to humans. Among the worst is the Crown-of-Thorns Starfish, which is covered with long, sharp, venomous spines. Its wounds are painful and the venom sometimes causes fever, vomiting, and temporary paralysis. Venoms of others cause a rash.

CROWN-OF-THORNS STARFISH
Acanthaster planci
to 20 in. (50 cm)
Indo-Pacific Ocean

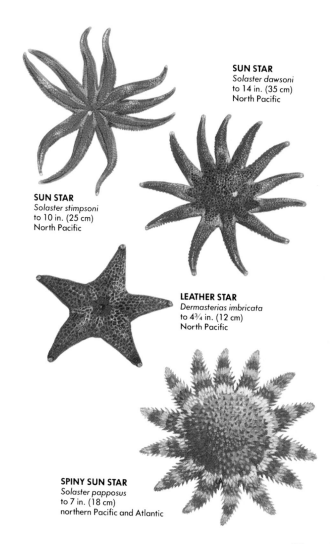

SUN STAR
Solaster dawsoni
to 14 in. (35 cm)
North Pacific

SUN STAR
Solaster stimpsoni
to 10 in. (25 cm)
North Pacific

LEATHER STAR
Dermasterias imbricata
to 4¾ in. (12 cm)
North Pacific

SPINY SUN STAR
Solaster papposus
to 7 in. (18 cm)
northern Pacific and Atlantic

25

SEA URCHINS, about 750 species, are generally globular and covered with spines. Scattered among the spines, too, are pedicellariae, which are like tiny sets of jaws on stalks. The tips of the jaws in some types are like fangs, and they are connected to venom glands. The primary function of the pedicellariae is to repel attackers.

In most species the spines are blunt and nonvenomous, but some have slender, brittle, hollow, sharp spines connected with venom glands. The venom can cause intense pain, and, in a few species, paralysis or even death.

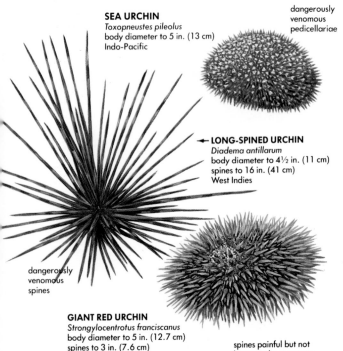

SEA URCHIN
Toxopneustes pileolus
body diameter to 5 in. (13 cm)
Indo-Pacific

dangerously
venomous
pedicellariae

←LONG-SPINED URCHIN
Diadema antillarum
body diameter to 4½ in. (11 cm)
spines to 16 in. (41 cm)
West Indies

dangerously
venomous
spines

GIANT RED URCHIN
Strongylocentrotus franciscanus
body diameter to 5 in. (12.7 cm)
spines to 3 in. (7.6 cm)
Pacific Coast of U.S.

spines painful but not
known to be venomous

SEA CUCUMBERS, about 500 species, have a sausagelike body and a ring of tentacles around their mouth. Some are brightly colored and produce toxins that repel predators. Toxins are located in the skin and in the tubules of Cuvier attached to the respiratory tree. When a sea cucumber is attacked, it may spew out its tubules of Cuvier in a sticky net. Some sea cucumbers are dangerously toxic if eaten. If the toxins get into the mouth or eyes, blindness or even death can result.

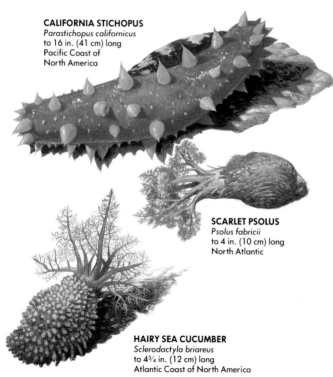

CALIFORNIA STICHOPUS
Parastichopus californicus
to 16 in. (41 cm) long
Pacific Coast of
North America

SCARLET PSOLUS
Psolus fabricii
to 4 in. (10 cm) long
North Atlantic

HAIRY SEA CUCUMBER
Sclerodactyla briareus
to 4¾ in. (12 cm) long
Atlantic Coast of North America

SPIDERS

About 37,000 species of spiders have been named so far. All spiders have a pair of fangs on their jaws (chelicerae), and ducts from a pair of venom glands located in the head lead to the hollow fangs. Spiders use the venom either to kill or to paralyze their prey. The venom of only a few species is dangerous to humans, but the venom of some 50 species in the United States and many more worldwide have caused pain and discomfort in humans.

Based on the arrangement of their jaws, spiders are divided into two groups: mygalomorphs, which have jaws attached at the front of the head and strike downward with their fangs; and true spiders, which have jaws attached below the head and strike sideways with their fangs that meet and cross in the middle.

Spiders can be kept out of homes and other buildings by getting rid of the insects on which they feed. An abnormal fear of spiders is called arachnophobia. Arachnid refers to spiders, scorpions, mites, ticks, and harvestmen—about 75,000 known species in all.

Fangs of a true spider (Black Widow), photographed with a scanning electron microscope

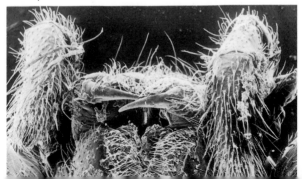

FUNNELWEB SPIDERS are mygalomorph spiders that entangle their prey in sheets of silk. A few species occur in Europe, and about ten species are found in the United States. None of these is known to be dangerous. Most species live in the tropics of the Southern Hemisphere, and some have potent venoms. The neurotoxic venom of some Australian species causes several human deaths per year, but nonfatal bites do not cause permanent nerve damage. Males are about five times more venomous than females and are especially aggressive. Some South American species also are aggressive and have caused human deaths.

**TREE-DWELLING
FUNNELWEB SPIDER**
Atrax formidabilis
to 2⅜ in. (60 mm)
southeastern Australia

female

SYDNEY FUNNELWEB SPIDER
Atrax robustus
to 1⅝ in. (40 mm)
southeastern Australia

male

TARANTULAS, which belong to a group known as bird spiders or hairy mygalomorphs, are the largest of all spiders. One South American species has a leg span of 10 inches. These spiders are found on all continents. About 30 species live in the United States, and while none of these are dangerous to humans, their bites are painful. Some from New Guinea, Australia, South America, and Africa are dangerously venomous, and occasionally a dangerous species enters the United States in fruit shipments. They are sometimes sold in pet shops, too. Their venom is essentially neurotoxic but also affects heart muscle and digests tissue. Tarantulas have sharp, barbed hairs on their abdomen. These urticating, or stinging hairs cause skin irritation and may give off toxic secretions. Many of these large spiders feed on small vertebrates, including mice, birds, frogs, lizards, and snakes—even poisonous species.

A tarantula's fangs, photographed with a scanning electron microscope

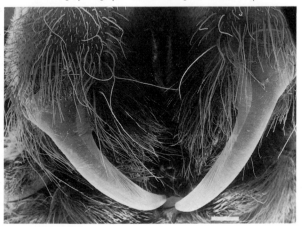

GUATEMALAN TARANTULA
Stichoplastus spinulosus
to 1⅝ in. (41 mm)
Guatemala

eating a shrew killed by venom

A tarantula's stinging hairs, photographed with a scanning electron microscope

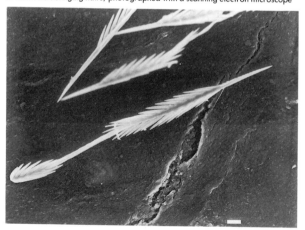

TARANTULAS

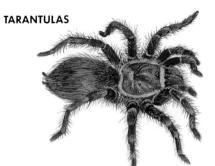

**TEXAS BROWN
TARANTULA**
Rhechostica hentzi
to 2 in. (50 mm)
southwestern U.S.

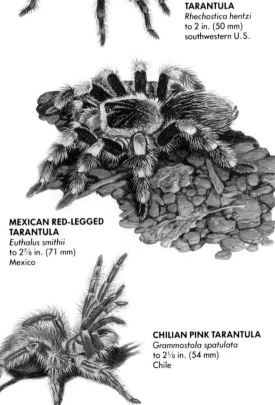

**MEXICAN RED-LEGGED
TARANTULA**
Euthalus smithii
to 2⅞ in. (71 mm)
Mexico

CHILIAN PINK TARANTULA
Grammostola spatulata
to 2⅛ in. (54 mm)
Chile

**CURLY-HAIRED
TARANTULA**
Euthalus albipilosa
to 2 in. (52 mm)
Central America

**PINK-TOED
BIRD-EATING SPIDER**
Avicularia avicularia
to 1⅜ in. (34 mm)
Amazon region

BROWN RECLUSE spiders belong to the true spider group, as do the following (pp. 35-37). Most true spiders (nearly 37,000 species) have fangs too short or venom too weak to injure humans. Only about 80 species worldwide are capable of injecting venom in humans.

The Brown Recluse, one of the two dangerous spiders in the United States, is most common in the southern states but may live in buildings throughout the country. The Brown Recluse and related species have been accidentally introduced around the world. The bites of many or all of these spiders initially burn and itch, or occasionally cause no distress at all. The area around the bite then turns red (or black and blue) as the tissue around the bite is digested away and falls off. The action of the venom of the Brown Recluse and related spiders is referred to as "loxoscelism," after the scientific name of these spiders. Another common name of the Brown Recluse is the Fiddleback because of the dark fiddle-shaped mark on the back.

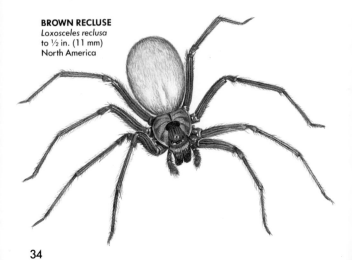

BROWN RECLUSE
Loxosceles reclusa
to ½ in. (11 mm)
North America

WIDOWS of six species occur nearly everywhere in the world but are most common in warm regions. Females of some of the widows have a venomous bite. Best known of these is the Black Widow, abundant in warm regions around human dwellings—especially wood piles, stone walls, and outside toilets. Most of the widows are shiny black with red markings on the abdomen. The Black Widow has a red hourglass-shaped mark on the belly. The Black Widow's neurotoxic venom often produces a painless bite that is followed later by cramps in the chest, abdomen, and muscles and sometimes by nausea, reduced heart rate, and shock. Full recovery can take a month or longer. About 5 people of every 100 bitten die. Poisoning by a widow's venom is called "latrodectism."

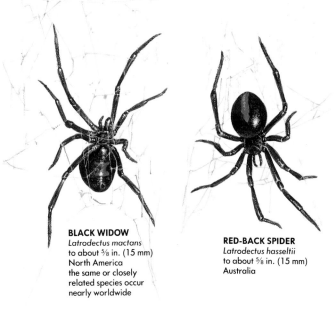

BLACK WIDOW
Latrodectus mactans
to about ⅝ in. (15 mm)
North America
the same or closely
related species occur
nearly worldwide

RED-BACK SPIDER
Latrodectus hasseltii
to about ⅝ in. (15 mm)
Australia

WOLF SPIDERS of about 3,000 species are largely confined to the Northern Hemisphere. About 200 species occur in the United States. None is dangerous, but the bites of some species cause localized pain, swelling, and light-headedness. The tarantella, a lively Italian folk dance, was believed to rid the body of a wolf spider's venom.

WOLF SPIDER
Trebacosa sp.
to ⅜ in. (9 mm)
Guatemala

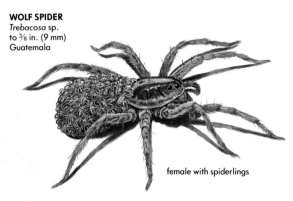

female with spiderlings

JUMPING SPIDERS are small and usually brightly colored. About 300 of the some 5,000 species occur in the United States. Some have been reported to give painful bites, with reaction to the venom lasting up to two weeks. None is dangerously venomous.

JUMPING SPIDER
Salticus scenicus
to about ½ in. (13 mm)
Europe and North America

DARING JUMPING SPIDER
Phidippus audax
to ⅝ in. (15 mm)
eastern two thirds of North America

WANDERING SPIDERS of some 550 species live in tropical regions. They hunt on the ground or in vegetation and do not build webs. The venom of some species is high in serotonin, which affects the nervous system and causes a very painful bite.

BANANA SPIDER
Cupiennius salei
to 1⅜ in. (36 mm)
Central and South America

GARDEN SPIDERS are large common spiders. Their bites cause pain, small open wounds, and symptoms suggestive of a neurotoxin, but they are probably not seriously dangerous. Garden spiders are members of the orb-weaver family containing about 3,500 species. Only about 180 species occur in the United States.

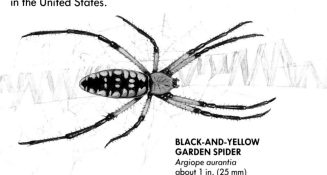

BLACK-AND-YELLOW GARDEN SPIDER
Argiope aurantia
about 1 in. (25 mm)
Oregon, California, and
eastern U.S.

CENTIPEDES

All members of one group of centipedes, the scolopendro-morphs (about 500 species), are venomous to some degree. These centipedes are nearly worldwide in distribution. Their first pair of "legs" is modified into hollow fangs (see p. 5), each with a venom gland at its base. Prey are seized and held with the fangs as the centipede feeds. Large centipedes—some to 12 inches long—may kill and feed on small birds, mammals, lizards, snakes, and frogs. All centipedes eat insects and other arthropods.

Centipede bites produce a burning pain, those from large species causing nausea and temporary paralysis. Deaths have occurred. Claw scratches of some species may cause pain due to toxins produced by glands in the walking legs. Some centipedes are colored bright red, orange, blue, green, or yellow; others are brown or black.

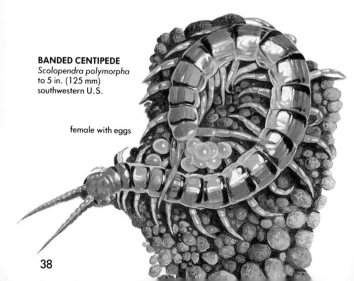

BANDED CENTIPEDE
Scolopendra polymorpha
to 5 in. (125 mm)
southwestern U.S.

female with eggs

GIANT NORTH AMERICAN CENTIPEDE
Scolopendra heros
to 7 in. (172 mm)
southern U.S.

BLUE-TAILED CENTIPEDE
Scolopendra sp.
to 2 in. (53 mm)
Texas

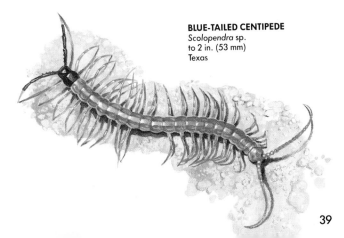

MILLIPEDES

Millipedes defend themselves by secreting toxins from glands either along the sides of the body or the middle of the back—or both. Many of the secretions are antibiotic, preventing growths of bacteria or fungi, and probably evolved for that purpose. Others are sticky when secreted but harden upon contact with the air. Their shell-like coating helps protect millipedes from ants and other small predators. The secretions of still other millipedes paralyze spiders, mice, and other creatures. More than 30 toxins have been identified from about 60 species. Among the toxins known in millipedes are hydrogen cyanide, formic acid, acetic acid, benzaldehyde, and phenol. Most of the world's about 8,000 species have not been examined. Millipedes are not known to be dangerous to humans.

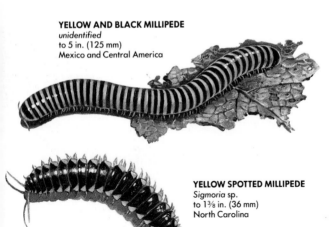

YELLOW AND BLACK MILLIPEDE
unidentified
to 5 in. (125 mm)
Mexico and Central America

YELLOW SPOTTED MILLIPEDE
Sigmoria sp.
to 1⅜ in. (36 mm)
North Carolina

WHIP SCORPIONS

Whip scorpions of about 85 species range throughout Central and South America, Asia, and into the southern United States. They do not sting or have a venomous bite. They do spray irritants from anal glands located at the base of the whip, however. The United States species sprays acetic acid that smells like vinegar, leading to the common name Vinegaroon. Other species spray formic acid or chlorine. The spray is not dangerous but is effective in repelling mice, birds, and other small predators.

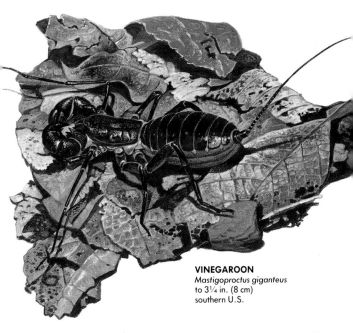

VINEGAROON
Mastigoproctus giganteus
to 3¼ in. (8 cm)
southern U.S.

SCORPIONS

Scorpions of roughly 1,200 species are nearly worldwide in distribution, but most species occur in dry, warm regions. A pair of venom glands is located in the stinger, which is the last segment of the tail. These glands are connected by ducts to the sharp tip of the stinger (see p. 7). When scorpions sting, muscles in the stinger cause venom to be injected into the wound. Scorpions use the venom to kill prey and to defend themselves.

Fewer than 50 species of scorpions are known to be dangerous to humans. The greatest risk from these species is to small children, as many as 50 percent of the stings causing fatalities. Scorpion stings are seldom fatal to healthy adults.

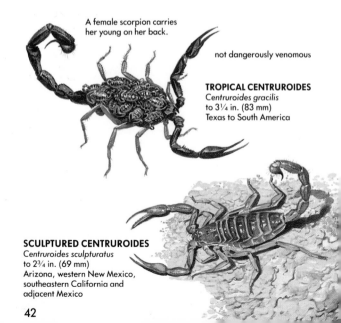

A female scorpion carries her young on her back.

not dangerously venomous

TROPICAL CENTRUROIDES
Centruroides gracilis
to 3¼ in. (83 mm)
Texas to South America

SCULPTURED CENTRUROIDES
Centruroides sculpturatus
to 2¾ in. (69 mm)
Arizona, western New Mexico,
southeastern California and
adjacent Mexico

42

SCULPTURED CENTRUROIDES is the only one of the 20 to 30 species of scorpions in the United States known to be dangerous to humans. Its sting causes severe pain, salivation, paralysis, and convulsions. Deaths are not common but do occur, especially in children. Other United States species, even members of the same genus, have painful stings but do not represent a serious threat.

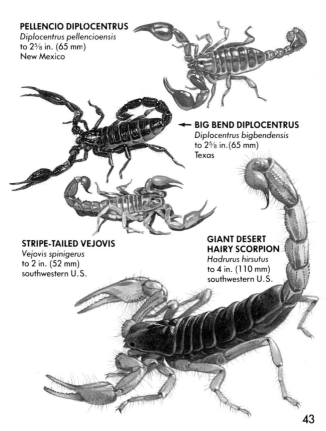

PELLENCIO DIPLOCENTRUS
Diplocentrus pellencioensis
to 2⅝ in. (65 mm)
New Mexico

← **BIG BEND DIPLOCENTRUS**
Diplocentrus bigbendensis
to 2⅝ in. (65 mm)
Texas

STRIPE-TAILED VEJOVIS
Vejovis spinigerus
to 2 in. (52 mm)
southwestern U.S.

GIANT DESERT HAIRY SCORPION
Hadrurus hirsutus
to 4 in. (110 mm)
southwestern U.S.

DANGEROUS SCORPIONS occur in North Africa and the Middle East, southern Africa, India, Mexico, and South America. Stings of these species cause intense local pain and swelling, followed by convulsions, paralysis, and sometimes death. Death may occur within only a few minutes or after several days. Typically, species with slender, weak pincers are dangerous; those with strong, heavy pincers have mild venoms that may cause intense pain but are not lethal.

EMPEROR SCORPION
Pandinis imperator
to 7 in. (180 mm)
Africa

Stings of the four scorpions shown here are painful but not life threatening.

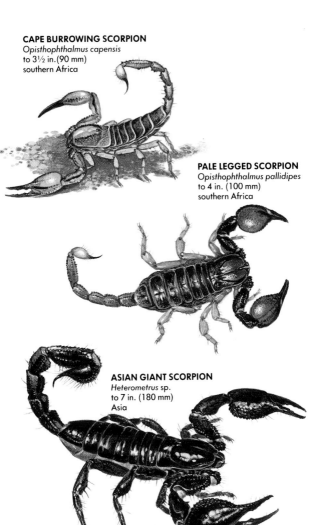

CAPE BURROWING SCORPION
Opisthophthalmus capensis
to 3½ in. (90 mm)
southern Africa

PALE LEGGED SCORPION
Opisthophthalmus pallidipes
to 4 in. (100 mm)
southern Africa

ASIAN GIANT SCORPION
Heterometrus sp.
to 7 in. (180 mm)
Asia

SCORPIONS

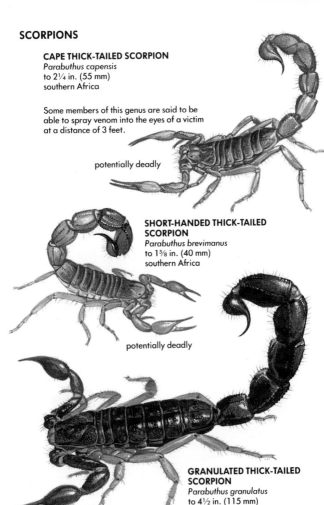

CAPE THICK-TAILED SCORPION
Parabuthus capensis
to 2¼ in. (55 mm)
southern Africa

Some members of this genus are said to be
able to spray venom into the eyes of a victim
at a distance of 3 feet.

potentially deadly

SHORT-HANDED THICK-TAILED SCORPION
Parabuthus brevimanus
to 1⅝ in. (40 mm)
southern Africa

potentially deadly

GRANULATED THICK-TAILED SCORPION
Parabuthus granulatus
to 4½ in. (115 mm)
southern Africa

potentially deadly

CARINATED THICK-TAILED SCORPION
Uroplectes carinatus
to 1⅝ in. (40 mm)
southern Africa

dangerous

HOUSE SCORPION
Euscorpius italicus
to 1⅝ in. (40 mm)
Mediterranean

painful but not dangerous

COMMON YELLOW SCORPION
Buthus occitanus
to 3 in. (75 mm)
Mediterranean and North Africa

potentially deadly

INSECTS

Nearly three fourths of all the known species of animals are insects. Insects have six legs, and most kinds have two pairs of wings. They either produce or retain from their food a bewildering array of chemical substances that are used to subdue prey or to repel predators. A number of insects have evolved resistance to the pesticides developed to control them. In a few cases, the insects even retain the chemicals and utilize them in their own defensive secretions.

Many insects, such as mosquitoes, black flies, horse flies, and some bugs, bite and feed on the blood of humans and other animals. Many of these species transmit diseases such as malaria, yellow fever, sleeping sickness, and Chagas' disease to humans. Saliva released in the bite wound sometimes causes an allergic reaction or mild pain. These insects are not treated in this book.

Other insects kill prey with a venomous saliva (primarily the true bugs, p. 49) or are equipped with stingers that inject venom from specialized glands at the rear of the body (bees, wasps, and ants, pp. 50-57). These venoms may also produce very painful reactions in humans.

Some insects repel predators by toxins. These chemicals may be associated with specially modified bristles (caterpillars and moths, pp. 58-59) or produced in glands that empty onto the surface of the body (beetles and bugs, p. 62). Insects that feed on poisonous plants normally concentrate the toxins from their food in their body tissues, and this makes them inedible to birds and other predators (butterflies, moths, beetles, and bugs, pp. 60-61).

A few insects not only produce toxic chemicals but also have special chambers from which these chemicals can be sprayed onto attackers (beetles, caterpillars, and earwigs, p. 63).

VENOMOUS SALIVA is produced by a number of insects, most notably the true bugs (hemipterans) that puncture their prey with a beak (proboscis) and then inject the saliva that both kills and then digests the animal. The bugs then suck out the digested contents of their kill. Primarily they eat other insects, but they will bite large animals, including humans, in defense. The larvae of some aquatic beetles feed in a similar manner and also produce painful bites in humans.

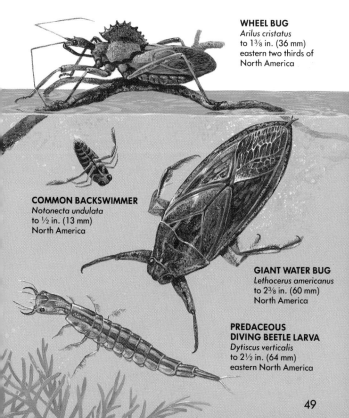

WHEEL BUG
Arilus cristatus
to 1⅜ in. (36 mm)
eastern two thirds of
North America

COMMON BACKSWIMMER
Notonecta undulata
to ½ in. (13 mm)
North America

GIANT WATER BUG
Lethocerus americanus
to 2⅜ in. (60 mm)
North America

**PREDACEOUS
DIVING BEETLE LARVA**
Dytiscus verticalis
to 2½ in. (64 mm)
eastern North America

STINGING INSECTS kill approximately 25 people per year in the United States. Most people who die are allergic to the venom, and death often occurs within a few minutes. Honeybees account for about 50 percent of the fatalities. Yellow Jackets and other wasps account for nearly all of the remaining deaths. Ant stings are rarely lethal. Only females have stingers, which are modified egg-laying structures (ovipositors). Social hymenopterans, such as Honeybees, sting in defense of their nest and are more likely to attack intruders than are solitary wasps that use their venom primarily to paralyze insect or spider prey to provision their nests. The wasp larvae eat the paralyzed prey during their development. Stinging insects seem to be attracted to strong perfumes and bright clothing.

HONEYBEES and their aggressive and more dangerous relative the African Honeybee (sometimes referred to as the Killer Bee) protect their hives by attacking intruders in number. The attackers are sterile females with barbed stingers that remain in the victim. The rear of the abdomen pulls out with the venom gland, which contracts—injecting its contents into the intruder. The nest is thereby protected even though the individual bees die. Honeybees are not native to North America but were introduced by early settlers. They live in colonies that may contain as many as 80,000 individuals.

HONEYBEE
Apis mellifera
worker to ½ in. (12 mm)
queen to ¾ in. (20 mm)
nearly worldwide

OTHER BEES native to North America are mostly solitary (except bumblebees). Each female provisions her own nest with pollen. Bumblebees live in much smaller colonies than do Honeybees. Native North American bees do not have barbed stingers and may therefore sting more than once.

RED-TAILED BUMBLEBEE
Bombus borealis
worker to ¾ in. (18 mm)
queen to ⅞ in. (23 mm)
east coast of northern U.S.
and Canada

GOLDEN NORTHERN BUMBLEBEE
Bombus fervidus
worker to ¾ in. (18 mm)
queen to ⅞ in. (23 mm)
North America

VIRESCENT GREEN METALLIC BEE
Agapostemon virescens
to ½ in. (12 mm)
eastern North America

FAITHFUL LEAF-CUTTING BEE
Megachile fidelis
to ½ in. (12 mm)
western half of U.S.

Honeybee stinger, photographed with a scanning electron microscope. Note the barbs.

SOCIAL WASPS, like Honeybees, attack in groups to defend their nests, but unlike Honeybees, they can sting repeatedly without dying. Their venoms are neurotoxic, consisting largely of histamine and serotonin. Their stings are very painful but are not dangerous unless a person is stung a great many times or is allergic.

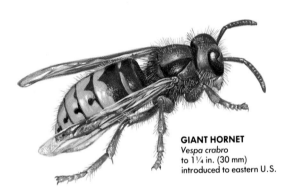

GIANT HORNET
Vespa crabro
to 1¼ in. (30 mm)
introduced to eastern U.S.

Stinger of a wasp, photographed with a scanning electron microscope

SANDHILLS HORNET
Vespula arenaria
to ¾ in. (20 mm)
northern U.S., Canada,
and Alaska

POLYBIINE PAPER WASP
Mischocyttarus flavitarsus
to ¾ in. (17 mm)
western North America

PAPER WASP
Polistes sp.
to 1 in. (25 mm)
Americas

SOLITARY WASPS are usually not aggressive, stinging only if attacked or captured. Many of these wasps are large and have powerful venom, but because of their mild temperaments, they are not dangerous to people.

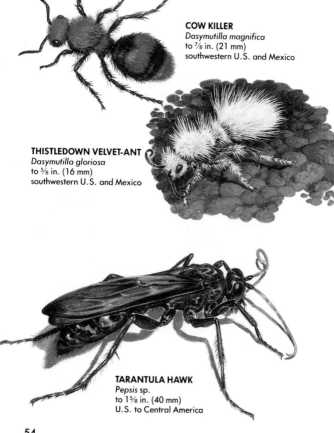

COW KILLER
Dasymutilla magnifica
to ⅞ in. (21 mm)
southwestern U.S. and Mexico

THISTLEDOWN VELVET-ANT
Dasymutilla gloriosa
to ⅝ in. (16 mm)
southwestern U.S. and Mexico

TARANTULA HAWK
Pepsis sp.
to 1⅝ in. (40 mm)
U.S. to Central America

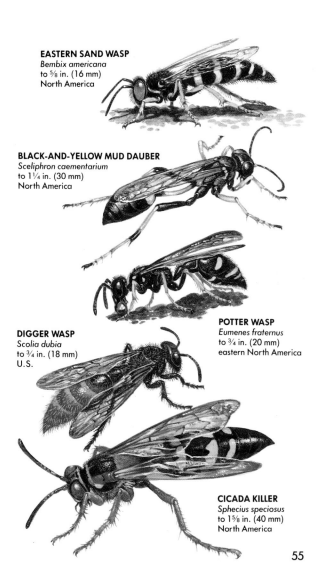

EASTERN SAND WASP
Bembix americana
to ⅝ in. (16 mm)
North America

BLACK-AND-YELLOW MUD DAUBER
Sceliphron caementarium
to 1¼ in. (30 mm)
North America

POTTER WASP
Eumenes fraternus
to ¾ in. (20 mm)
eastern North America

DIGGER WASP
Scolia dubia
to ¾ in. (18 mm)
U.S.

CICADA KILLER
Sphecius speciosus
to 1⅝ in. (40 mm)
North America

55

ANTS are flightless relatives of bees and wasps. Ants are found throughout the world, with most species occurring in the tropics. Reproductive ants have wings but shed them after they mate and swarm. The primary venom of ants is formic acid, which blocks the respiratory mechanism of other insects and is very painful to people and other vertebrates.

In the United States, the most aggressive and dangerous are fire ants and harvester ants. An estimated 2½ million people are stung each month in the United States by fire ants. Many of these tiny ants will crawl onto an intruder, and the first ant to sting releases a chemical signaling all the other ants to sting. Introduced accidentally to the United States from Brazil, the Tropical Fire Ant makes mounds 2 feet in diameter and as much as 3 feet high. Hundreds of thousands of ants live in a single mound.

TROPICAL FIRE ANT
Solenopsis geminata
to ¼ in. (6 mm)
southwestern U.S. and
Pacific Coast

A Tropical Fire Ant, photographed with a scanning electron microscope

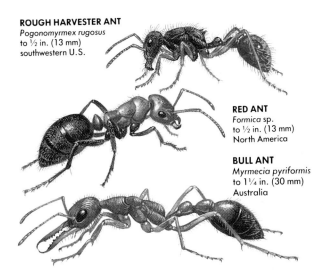

ROUGH HARVESTER ANT
Pogonomyrmex rugosus
to ½ in. (13 mm)
southwestern U.S.

RED ANT
Formica sp.
to ½ in. (13 mm)
North America

BULL ANT
Myrmecia pyriformis
to 1¼ in. (30 mm)
Australia

ACACIA ANTS live in the Americas and in Africa where they nest in hollow growths produced by the acacias. These highly aggressive ants inflict very painful stings that protect the acacias from animals that would otherwise eat the leaves.

American Ant Acacia

African Ant Acacia

STINGING HAIRS occur on hundreds (perhaps even thousands) of species of moths and butterflies. In most they are found only in the larval or caterpillar stages, but a few adult moths are protected by stinging scales. As in spiders, they cause a rash. In caterpillars, these urticating hairs are stiff and sharp, and some are barbed and hollow, with venom transmitted from sacs at their base. Little is known about the venoms involved, but some consist of a histaminelike substance or formic acid. Both compounds produce a burning pain when introduced into the skin. Common symptoms are fever, inflammation, and nausea, but a few people have died as a result of caterpillar stings, probably due to allergic reactions. Many hairy caterpillars like the ones illustrated here are very painful to touch, but other species that appear to have spines are harmless.

These caterpillars have spiky hairs that are embedded in victims with a rapid thrashing motion of the front half of the body.

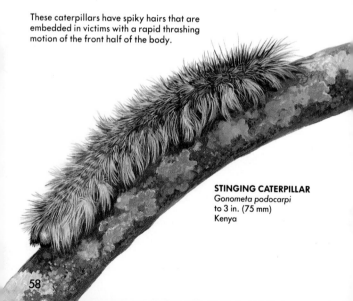

STINGING CATERPILLAR
Gonometa podocarpi
to 3 in. (75 mm)
Kenya

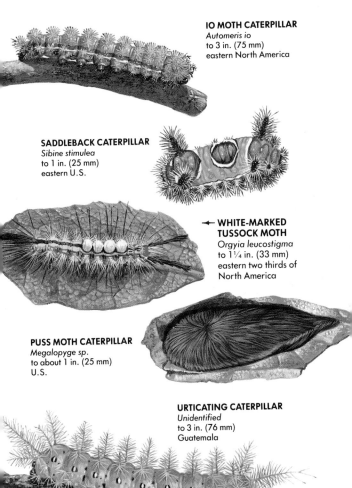

IO MOTH CATERPILLAR
Automeris io
to 3 in. (75 mm)
eastern North America

SADDLEBACK CATERPILLAR
Sibine stimulea
to 1 in. (25 mm)
eastern U.S.

← **WHITE-MARKED TUSSOCK MOTH**
Orgyia leucostigma
to 1¼ in. (33 mm)
eastern two thirds of
North America

PUSS MOTH CATERPILLAR
Megalopyge sp.
to about 1 in. (25 mm)
U.S.

URTICATING CATERPILLAR
Unidentified
to 3 in. (76 mm)
Guatemala

DISTASTEFUL AND POISONOUS INSECTS Many plants produce powerful toxins that prevent them from being eaten. This is generally effective against mammals, but many insects have developed a resistance to the toxins. They eat the plants, and the toxins concentrate in their tissues. As a result, these insects become poisonous to predators. Many have bright coloration that advertises their distastefulness.

SMALL WHIRLIGIG BEETLE
Gyrinus sp.
to ¼ in. (7 mm)
North America

RATTLEBOX MOTH
Utetheisa bella
to 1¾ in. (46 mm)
North America

COLORADO POTATO BEETLE
Leptinotarsa decimlineata
to ½ in. (11 mm)
North America

Toxins also act to attract mates.

CINNABAR MOTH CATERPILLAR
Callimorpha jacobaeae
to 1¼ in. (31 mm)
introduced into U.S. and
Canada from Europe

Caterpillars accumulate alkaloids
from the ragwort plant (*Senecio
jacobaeae*).

GOLDEN NET-WING BEETLE
Lycostomus loripes
to ⅜ in. (9 mm)
Arizona

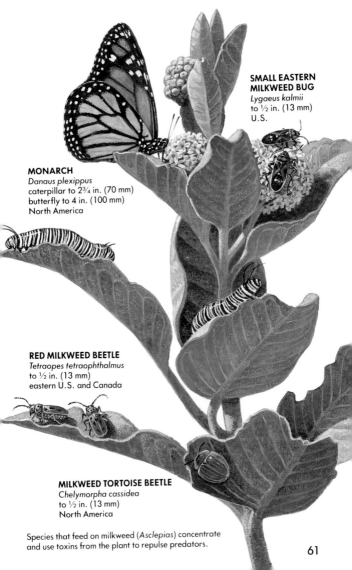

SMALL EASTERN MILKWEED BUG
Lygaeus kalmii
to ½ in. (13 mm)
U.S.

MONARCH
Danaus plexippus
caterpillar to 2¾ in. (70 mm)
butterfly to 4 in. (100 mm)
North America

RED MILKWEED BEETLE
Tetraopes tetraophthalmus
to ½ in. (13 mm)
eastern U.S. and Canada

MILKWEED TORTOISE BEETLE
Chelymorpha cassidea
to ½ in. (13 mm)
North America

Species that feed on milkweed (*Asclepias*) concentrate
and use toxins from the plant to repulse predators.

POISONOUS BODY FLUIDS and glandular secretions produced by many insects repel predators. They vary from the odor of stinkbugs to the irritating fluids of blister beetles. These substances may be dangerous to humans if eaten. This is especially true of the cantharidin produced by blister beetles. The chemical severely irritates the urinary tract. There are reports of people even being affected after eating frogs that had made meals of these beetles.

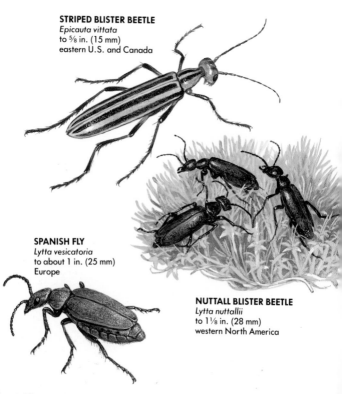

STRIPED BLISTER BEETLE
Epicauta vittata
to ⅝ in. (15 mm)
eastern U.S. and Canada

SPANISH FLY
Lytta vesicatoria
to about 1 in. (25 mm)
Europe

NUTTALL BLISTER BEETLE
Lytta nuttallii
to 1⅛ in. (28 mm)
western North America

INSECTS THAT SPRAY TOXINS use these chemicals to repel attackers. The compounds usually irritate the mouth lining and eyes, often causing temporary blindness. Bombardier Beetles spray boiling hot chemicals (quinones) produced by an explosion in a specialized chamber at the rear of the body. These sprays erupt with an audible popping sound and can blister the skin of humans and small predators.

Some darkling beetles spray quinones with small explosions and posture in a characteristic head-stand while spraying. Other beetles and some ants spray formic acids.

Caterpillar ejects formic acid from filaments on abdomen.

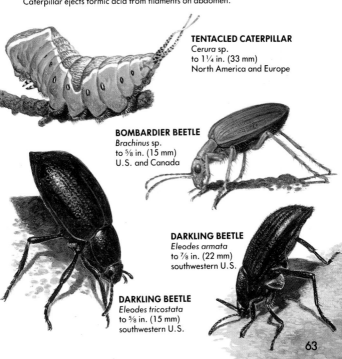

TENTACLED CATERPILLAR
Cerura sp.
to 1¼ in. (33 mm)
North America and Europe

BOMBARDIER BEETLE
Brachinus sp.
to ⅝ in. (15 mm)
U.S. and Canada

DARKLING BEETLE
Eleodes armata
to ⅞ in. (22 mm)
southwestern U.S.

DARKLING BEETLE
Eleodes tricostata
to ⅝ in. (15 mm)
southwestern U.S.

MIMICRY is a protection derived from resembling another organism—the mimic looking like the model. A predator cannot distinguish one from the other. Mimicry is quite common in insects. There are three common forms: (1) Special Resemblance, in which an insect looks like a thorn, stick, rock, or some other inedible object; (2) Batesian Mimicry, in which an edible insect resembles an inedible insect; and (3) Mullerian Mimicry, in which two inedible species resemble each other, such as the yellow-and-black color patterns of wasps.

ANT-MIMIC SPIDER
Castianeira sp.
to ⅜ in. (10 mm)
southeastern U.S.

ANT-MIMIC JUMPING SPIDER
Peckhamia picata
to ¼ in. (5 mm)
eastern U.S. and Canada

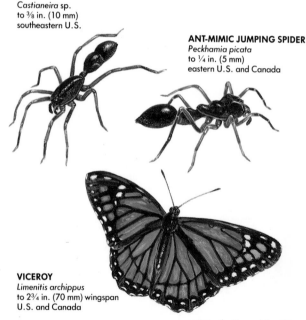

VICEROY
Limenitis archippus
to 2¾ in. (70 mm) wingspan
U.S. and Canada

mimics the Monarch (p. 61)

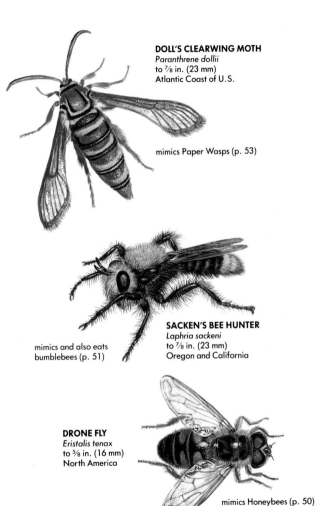

DOLL'S CLEARWING MOTH
Paranthrene dollii
to ⅞ in. (23 mm)
Atlantic Coast of U.S.

mimics Paper Wasps (p. 53)

SACKEN'S BEE HUNTER
Laphria sackeni
to ⅞ in. (23 mm)
Oregon and California

mimics and also eats
bumblebees (p. 51)

DRONE FLY
Eristalis tenax
to ⅝ in. (16 mm)
North America

mimics Honeybees (p. 50)

65

FISHES

More than half of all the world's animals with backbones (vertebrates) are fishes, totaling more than 21,500 species. They belong to four distinctly different vertebrate classes. A large number of fishes produce toxins or are venomous, but only a small percentage are covered in this book.

LAMPREYS (Class Cephalaspidomorphi, 36 species) and Hagfishes (Class Myxini, 32 species) are eel-like and jawless. Both secrete toxic skin slime. Also, the uncooked blood of these and various true eels (bony fish, p. 72), such as common, conger, moray and snake eels, is toxic. The blood has a burning bitter taste, irritates the eyes, and can cause death if eaten. The toxin prevents coagulation of blood and causes internal bleeding.

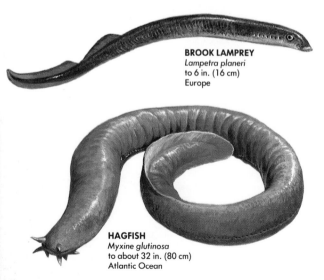

BROOK LAMPREY
Lampetra planeri
to 6 in. (16 cm)
Europe

HAGFISH
Myxine glutinosa
to about 32 in. (80 cm)
Atlantic Ocean

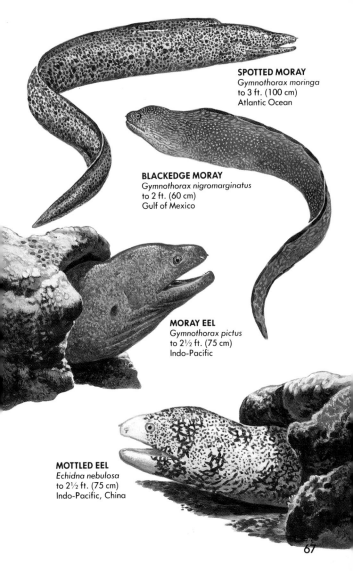

SPOTTED MORAY
Gymnothorax moringa
to 3 ft. (100 cm)
Atlantic Ocean

BLACKEDGE MORAY
Gymnothorax nigromarginatus
to 2 ft. (60 cm)
Gulf of Mexico

MORAY EEL
Gymnothorax pictus
to 2½ ft. (75 cm)
Indo-Pacific

MOTTLED EEL
Echidna nebulosa
to 2½ ft. (75 cm)
Indo-Pacific, China

RAYS, skates, sharks, and chimaeras are members of the Class Chondrichthyes, and they have skeletons of cartilage rather than bone. Many species have venomous spines, and many also have poisonous livers. A few have poisonous flesh. The roughly 320 species of rays and skates occur in seas throughout the world. Two species live in fresh water in South America.

Stingrays are a hazard to people wading in shallow, warm seas. In the United States alone, about 750 people are stung every year, but very few deaths result. Stingrays have a serrated (saw-toothed) spine on the upper surface of the tail, and when the tail is lashed, this spine cuts into the victim. A glandular skin covering the spine or sting produces a venom, and some of the venom remains in the wound when the spine is withdrawn. Stings cause intense pain followed by loss of blood pressure and an erratic heartbeat. Largest is the Indo-Pacific Smooth Stingray—up to 7 feet wide and 15 feet long, with a sting to as much as 1 foot long. Some electric rays are as much as 6 feet long and weigh over 200 pounds. While not venomous, they can generate shocks as great as 220 volts—enough to stun an adult human.

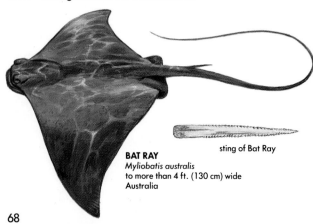

sting of Bat Ray

BAT RAY
Myliobatis australis
to more than 4 ft. (130 cm) wide
Australia

FRESHWATER STINGRAY
Potamotrygon reticulatus
to 20 in. (50 cm) wide
South America

sting of Freshwater Stingray

YELLOW STINGRAY
Urolophus jamaicensis
to 2 ft. 3 in. (67 cm) long
Caribbean

sting of Yellow Stingray

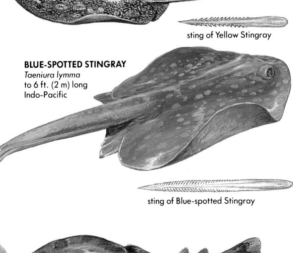

BLUE-SPOTTED STINGRAY
Taeniura lymma
to 6 ft. (2 m) long
Indo-Pacific

sting of Blue-spotted Stingray

weakly electric

LESSER ELECTRIC RAY
Narcine brasiliensis
to 18 in. (46 cm) long
Atlantic, U.S. to South America

SHARKS of about 370 species occur in oceans throughout the world. Some have venom glands in the skin covering the spines near their fins. The occasional attacks of sharks on swimmers represent a greater danger to humans than do those with venom, however.

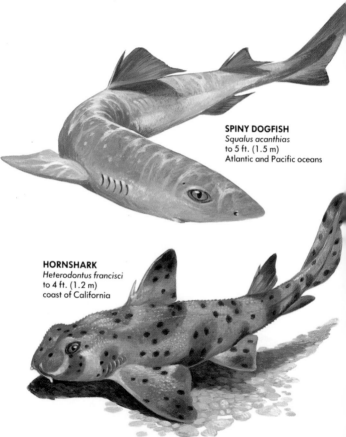

SPINY DOGFISH
Squalus acanthias
to 5 ft. (1.5 m)
Atlantic and Pacific oceans

HORNSHARK
Heterodontus francisci
to 4 ft. (1.2 m)
coast of California

CHIMAERAS, also known as ratfishes, are odd-looking members of the Class Chondrichthyes. All of the 23 species are marine and are most common in temperate waters. Venom from glands associated with at least the dorsal fin spines can cause extremely painful puncture wounds. Ratfishes can also give severe but not venomous bites.

RATFISH
Hydrolagus colliei
to 3 ft. (1 m)
Pacific Coast of North America

dorsal spine of Ratfish

BONY FISHES (Class Osteichthyes) of some 20,750 species are common throughout the world in both fresh and salt waters. Many species have venomous spines, others have toxic skin secretions, and at least one has a venomous bite.

CATFISHES of about 2,400 species occur nearly throughout the world in fresh waters and tropical oceans. Most of them have venomous spines, and the spines, at least on the pectoral and dorsal fins, are usually toothed, or serrated. Many species also are heavily armored with spines along the body. The fin spines and dorsal spines of many species can be locked in an erect position. Also, the spines of some species are hollow and inject venom into wounds. Stings are painful and cause deadening, but most species are not known to be serious health hazards to humans.

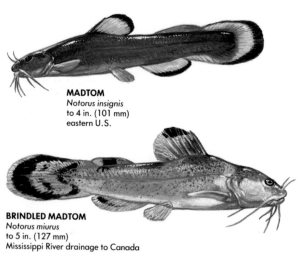

MADTOM
Notorus insignis
to 4 in. (101 mm)
eastern U.S.

BRINDLED MADTOM
Notorus miurus
to 5 in. (127 mm)
Mississippi River drainage to Canada

These and other members of the genus *Notorus* in the U.S. give painful but not dangerous puncture wounds with their spines.

SOUTH AMERICAN CATFISHES

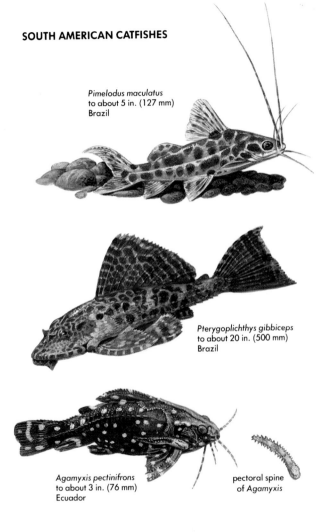

Pimelodus maculatus
to about 5 in. (127 mm)
Brazil

Pterygoplichthys gibbiceps
to about 20 in. (500 mm)
Brazil

Agamyxis pectinifrons
to about 3 in. (76 mm)
Ecuador

pectoral spine
of *Agamyxis*

AFRICAN CATFISHES

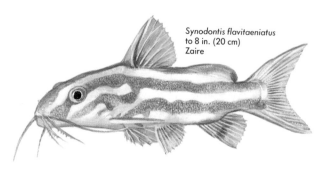

Synodontis flavitaeniatus
to 8 in. (20 cm)
Zaire

Mochokiella paynei
to 4 in. (10 cm)
West Africa

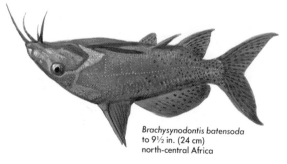

Brachysynodontis batensoda
to 9½ in. (24 cm)
north-central Africa

MARINE CATFISH are among the most venomous of all fishes. Glands supply both the dorsal and pectoral spines with venom that is both neurotoxic and hemotoxic. The potent toxin has been responsible for human deaths.

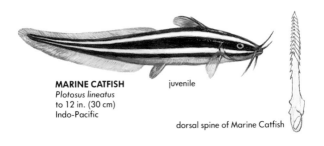

MARINE CATFISH
Plotosus lineatus
to 12 in. (30 cm)
Indo-Pacific

juvenile

dorsal spine of Marine Catfish

INDIAN CATFISHES are also deadly venomous and allegedly attack and sting people in the water.

INDIAN CATFISH
Heteropneustes fossilis
to 10 in. (25 cm)
India to Vietnam

SCORPIONFISHES and their relatives (about 1,000 species) occur in shallow marine waters throughout the world. The venom is produced by glands along the spine or at their base, and its strength varies greatly. Some have grooved venom spines located on the head and all fins. Scorpionfishes often posture the body and jab spines into their victims. The stings cause severe pain or death. The Stone Fish from Australia has the largest venom gland of any fish and is recognized as such a hazard that an antivenin has been developed.

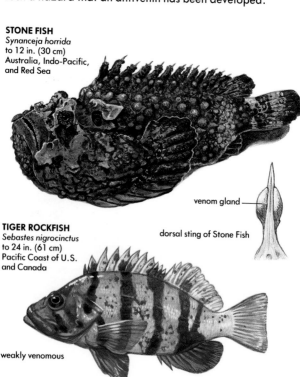

STONE FISH
Synanceja horrida
to 12 in. (30 cm)
Australia, Indo-Pacific,
and Red Sea

venom gland

dorsal sting of Stone Fish

TIGER ROCKFISH
Sebastes nigrocinctus
to 24 in. (61 cm)
Pacific Coast of U.S.
and Canada

weakly venomous

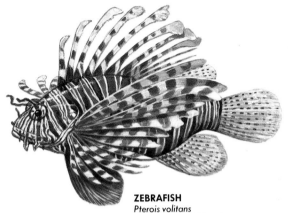

ZEBRAFISH
Pterois volitans
to about 12 in. (30 cm)
Indo-Pacific, China, Australia,
and Red Sea

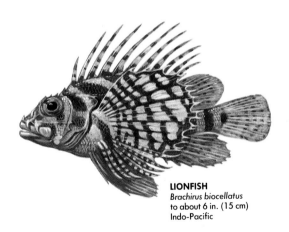

LIONFISH
Brachirus biocellatus
to about 6 in. (15 cm)
Indo-Pacific

WEEVERFISHES occur along the eastern Atlantic and Mediterranean coasts. Venomous stings result from contact with the five or more dorsal spines or the dagger-shaped spine on the gill cover (operculum) on each side of the head. Victims suffer excruciating pain, often screaming and thrashing about before losing consciousness. Stings are occasionally fatal.

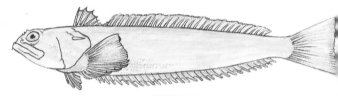

GREATER WEEVERFISH
Trachinus draco
to 18 in. (45 cm)
northeastern Atlantic,
Mediterranean, North Africa

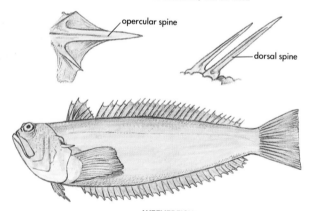

opercular spine

dorsal spine

WEEVERFISH
Trachinus radiatus
to 10 in. (25 cm)
Mediterranean, west coast of Africa

TOADFISHES of about 60 species occur nearly worldwide in marine waters, a few in fresh water. They are bottom dwellers with broad flat heads and large mouths. Toadfishes erect their dorsal and opercular spines aggressively when disturbed. The highly developed venom mechanism consists of a hollow spine through which venom is injected as the gland is pressed when the spine pierces a victim. Wounds are usually the result of a toadfish being stepped on while it is buried in the mud. The stings cause great pain, and the swelling may last for months. No fatalities have been reported.

TOADFISH
Opsanus tau
to 15 in. (38 cm)
Atlantic Coast of U.S.

STARGAZERS of about ten species live in tropical areas. They have spines (cleithral) on each shoulder, and each is surrounded by a venom gland. Little is known about the effects of the venom.

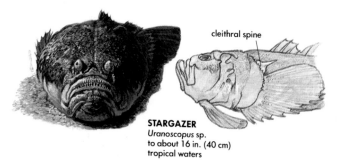

cleithral spine

STARGAZER
Uranoscopus sp.
to about 16 in. (40 cm)
tropical waters

SURGEONFISHES (about 1,000 species) inhabit tropical seas. Erectable spines at the base of the tail can inflict painful wounds. When a spine is drawn back into its sheath, or pocket, it is covered with venom produced by the pocket lining.

SURGEONFISH
Acanthurus dussumieri
to 12 in. (30 cm)
South Pacific

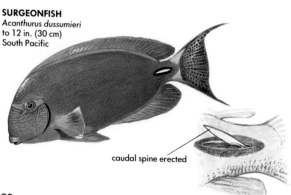

caudal spine erected

RED SEA SOLES and Pacific Soles secrete from skin glands toxins known as pavoninins. They are potent enough to repel predatory fishes, including sharks.

RED SEA SOLE
Pardachirus marmoratus
to 10 in. (25 cm)
Red Sea

BLENNIES of the genus *Meiacanthus* are the only fishes known to have fangs and venomous bites. When the fish bites, the milky venom flows up the deeply grooved fangs in the lower jaw. The venom is not used when the fish is feeding but is released when the fish bites in its own defense.

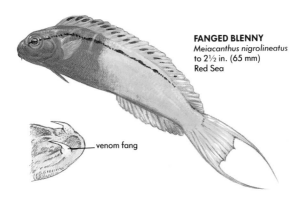

FANGED BLENNY
Meiacanthus nigrolineatus
to 2½ in. (65 mm)
Red Sea

venom fang

TETRODOTOXIN is one of the strongest toxins known. A neurotoxin, it is found in salamanders (pp. 98-101), frogs (p. 92), octopuses (p. 21), snails (p.18), and fish in at least four families. They are known as puffers or porcupine fishes. Tetrodotoxin is found also in the Ocean Sunfish and probably seven other families of related fishes, including triggerfishes, filefishes, spikefishes, and trunkfishes. Both marine and fresh-water species are toxic. The greatest concentrations of the toxin are in the skin, liver, and ovaries, but even small amounts of muscle tissue from some species can be lethal. Many of these fishes have sharp, spine-shaped scales surrounded by toxic glandular skin.

Puffers are eaten in Japan as raw "sashimi fugu" and in a soup called "chiri." Most cases of poisoning are from the soup. Eating sashimi fugu often causes intoxication, with light-headedness and numbness of the lips. It is eaten by many, in fact, to get these effects.

Fatal doses can cause death from within a few minutes to a day. If a victim survives longer than 24 hours, recovery is usual. Apparently, a victim can seem to be comatose yet remain conscious and mentally alert. Some have recovered after several days in an apparent coma and have claimed to remember everything that happened.

The first symptoms of a dangerous dosage include deadening of the lips and tongue, dizziness, and vomiting within a few minutes. These are followed by numbness and prickling over the entire body, rapid heartbeat, decreased blood pressure, and paralysis of muscles. Death is caused by suffocation when the diaphragm muscles are paralyzed and breathing stops.

Tetrodotoxin has many uses. It is isolated from puffer fishes in Japan. The purified toxin has been used to reduce pain and in research on the nervous system. Puffer fishes are reputedly used in voodoo to turn victims into "zombies."

at rest

inflated

PUFFER
Sphaeroides maculatus
to 10 in. (25 cm)
Atlantic, both North and
Central America

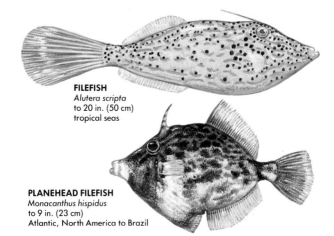

FILEFISH
Alutera scripta
to 20 in. (50 cm)
tropical seas

PLANEHEAD FILEFISH
Monacanthus hispidus
to 9 in. (23 cm)
Atlantic, North America to Brazil

83

TETRODOTOXIC FISHES

PORCUPINEFISH
Chilomycterus orbicularis
to 10 in. (25 cm)
Indo-Pacific

BALLOONFISH
Diodon holacanthus
to 18 in. (46 cm)
tropics worldwide

PORCUPINEFISH
Diodon hystrix
to 3 ft. (91 cm)
Atlantic and Pacific

inflated

PUFFER
Arothron nigropunctatus
to 10 in. (25 cm)
Indo-Pacific

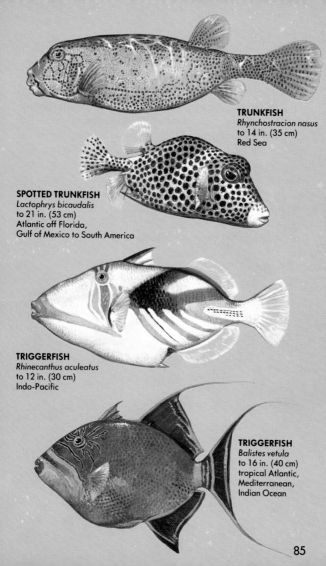

TRUNKFISH
Rhynchostracion nasus
to 14 in. (35 cm)
Red Sea

SPOTTED TRUNKFISH
Lactophrys bicaudalis
to 21 in. (53 cm)
Atlantic off Florida,
Gulf of Mexico to South America

TRIGGERFISH
Rhinecanthus aculeatus
to 12 in. (30 cm)
Indo-Pacific

TRIGGERFISH
Balistes vetula
to 16 in. (40 cm)
tropical Atlantic,
Mediterranean,
Indian Ocean

85

AMPHIBIANS

Amphibians (frogs, salamanders, and caecilians—about 4,000 species in all) have a smooth, glandular skin. Mucous glands keep the skin moist. Granular glands, which are found in the skin in many species, produce toxins that repel predators and may also prevent infection by microorganisms. The toxins are not used to kill or subdue prey. In some species the toxins produced by the skin glands are deposited in the eggs and protect the developing embryos. The larvae, which are normally aquatic, are unprotected by glandular secretions until their own skin glands become active when they metamorphose into their adult terrestrial stage.

More than 200 different kinds of toxins produced by amphibians have been described, and many more await

AMERICAN TOAD
Bufo americanus
to 4½ in. (118 mm)
eastern U.S. and Canada

The male may cling to the female for days while she lays eggs in long, gelatinous strings. American Toads lay as many as 25,000 toxic eggs in one breeding period.

86

identification. The toxin-producing glands are usually enlarged and often are concentrated in specific regions of the body. Amphibians posture their body so that an attacking predator will make contact with concentrations of the toxic, distasteful secretions. Some snakes have evolved a resistance to amphibian toxins.

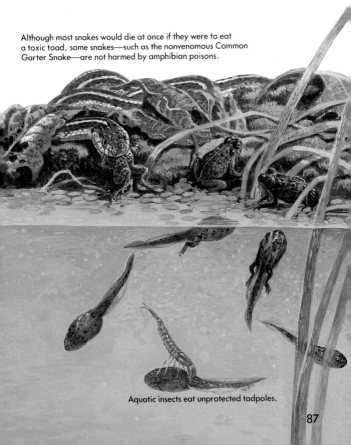

Although most snakes would die at once if they were to eat a toxic toad, some snakes—such as the nonvenomous Common Garter Snake—are not harmed by amphibian poisons.

Aquatic insects eat unprotected tadpoles.

TRUE TOADS (genus *Bufo*) occur worldwide except in Australia and the deserts of North Africa. Toads are frogs of the family Bufonidae. In their skin glands they produce a number of secretions that repel predators. These secretions, which include bufotoxin, bufotenin, and adrenalin, are concentrated in the parotoid glands, one on each side of the neck. If eaten or if applied to scratched skin, these toxins stimulate the heart rate, act as local irritants, and cause numbing.

Only two of the 17 species of true toads in the United States are dangerous to handle: the Marine Toad and the Colorado River Toad. Human deaths have been caused by eating Marine Toad eggs. In Hawaii, an estimated 50 dogs die every year as a result of biting Marine Toads. The Colorado River Toad produces the highly hallucinogenic toxin O-methyl-bufotenin.

WESTERN TOAD
Bufo boreas
to 5 in. (127 mm)
western U.S. and Canada

toxic glandular secretions

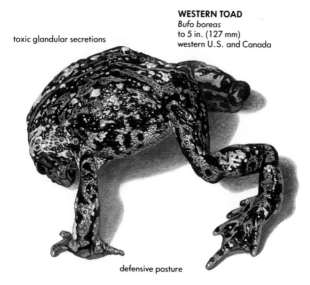

defensive posture

MARINE TOAD
Bufo marinus
to 9 in. (229 mm)
southern Texas to South America

parotoid glands

The parotoids and other wartlike skin glands
produce a foul-tasting poison that discourages
predators.

parotoid glands

COLORADO RIVER TOAD
Bufo alvarius
to 6 in. (150 mm)
southern Arizona, adjacent areas

ALL TRUE TOADS have venomous skin secretions that provide effective protection from birds, foxes, and most other small predators. The United States toads on these pages are not dangerous to people. Warts on toads are groups of enlarged skin glands. Toads do *not* cause warts in humans!

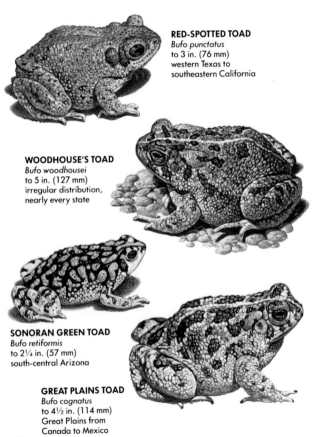

RED-SPOTTED TOAD
Bufo punctatus
to 3 in. (76 mm)
western Texas to
southeastern California

WOODHOUSE'S TOAD
Bufo woodhousei
to 5 in. (127 mm)
irregular distribution,
nearly every state

SONORAN GREEN TOAD
Bufo retiformis
to 2¼ in. (57 mm)
south-central Arizona

GREAT PLAINS TOAD
Bufo cognatus
to 4½ in. (114 mm)
Great Plains from
Canada to Mexico

MOST FROGS in the United States are safe to handle. They do produce a variety of skin secretions, including histamine and serotonin, that repel predators. In humans, these secretions will cause a burning sensation if they get into the eyes or in open scratches in the skin.

GREATER GRAY TREEFROG
Hyla versicolor
to 2⅜ in. (60 mm)
most of eastern U.S.

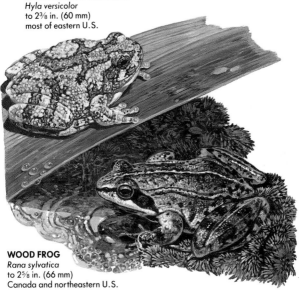

WOOD FROG
Rana sylvatica
to 2⅝ in. (66 mm)
Canada and northeastern U.S.

COUCH'S SPADEFOOT
Scaphiopus couchi
to 3½ in. (90 mm)
western Texas and
desert Southwest

FROGS from all over the world may have toxins produced by skin glands. Many of these secretions are painful and dangerous in an open scratch or cut.

SOUTH AMERICAN BULLFROG
Leptodactylus pentadactylus
to 8½ in. (216 mm)
South and Central America

HARLEQUIN FROG
Atelopus varius
to 2⅜ in. (58 mm)
Costa Rica to Colombia

Eggs of this frog are protected by tetrodotoxin (p. 82).

DARWIN'S FROG
Rhinoderma darwinii
to 1¼ in. (30 mm)
Chile and Argentina

Skin of Darwin's Frog has highest known levels of serotonin, a chemical that at high doses affects the nervous system.

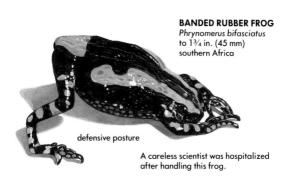

BANDED RUBBER FROG
Phrynomerus bifasciatus
to 1¾ in. (45 mm)
southern Africa

defensive posture

A careless scientist was hospitalized
after handling this frog.

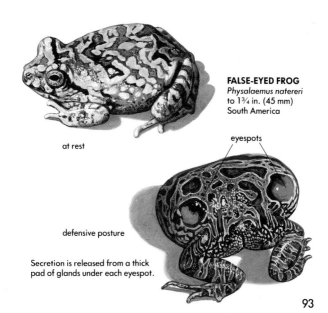

FALSE-EYED FROG
Physalaemus natereri
to 1¾ in. (45 mm)
South America

at rest

eyespots

defensive posture

Secretion is released from a thick
pad of glands under each eyespot.

93

POISON DART FROGS are small, brightly colored frogs of Central and South America. They produce a variety of cardiac and nerve toxins, more than 200 of them (steroidal alkaloids) described. These frogs are sufficiently distasteful and toxic to repel all predators except a few kinds of snakes. The most toxic of these frogs is *Phyllobates terribilis* from Colombia. It contains enough toxin to kill 20,000 mice and is dangerous to handle. Some South American Indians use the skin secretions to poison the tips of their blowgun darts.

Top left:
Dendrobates pumilio
to 1 in. (24 mm)
Nicaragua to Panama

Top right:
Dendrobates auratus
to 1½ in. (39 mm)
Nicaragua to Colombia

Bottom left:
Dendrobates lehmanni
to 1⅜ in. (35 mm)
Colombia

Bottom right:
Dendrobates granuliferus
to ⅞ in. (22 mm)
Costa Rica

Indian using blowgun

Dart being coated with skin toxins of a Poison Dart Frog

Phyllobates terribilis
to 1⅞ in. (47 mm)
Colombia

95

THE UNKEN REFLEX is an unusual behavior pattern exhibited by some of the most toxic frogs. Though they behave similarly, the frogs are not closely related. When attacked by predators, the frogs shut their eyes and bend their head and legs back over their body so that the bright colors on their bellies and the bottoms of their feet are exposed. This curved posture and color display presumably warn a predator of danger, sparing both the frog and its attacker.

Best known for this behavior is the European Fire-bellied Toad, called Unke in German, and thus the term "unken." Many newts (pp. 98-102) also exhibit this behavior pattern.

YELLOW-BELLIED TOAD
Bombina variegata
to 2 in. (50 mm)
Europe

at rest

unken reflex

Melanophryniscus stelzneri
to 1¼ in. (30 mm)
Brazil to Argentina

at rest

unken reflex

HARLEQUIN FROG
Atelopus pulcher
to 1⅜ in. (35 mm)
Ecuador

at rest

unken reflex

NEWTS are rough-skinned salamanders of the family Salamandridae. The skin glands of many (perhaps all) newts produce tetrodotoxin and other as yet unidentified toxins. The secretions burn the eyes and mouth and prevent the newts from being eaten by all predators except some garter snakes. Both human and animal deaths have resulted from eating western newts (genus *Taricha*). Newts are not dangerous to handle unless there are open cuts in the skin.

Scientists have determined the toxicity of some newts in mouse units (the amount of toxin needed to kill a 20-gram mouse in 10 minutes). Western newts have as much as 25,000 mouse units of toxin. Efts (terrestrial juveniles) of the Eastern Newt are less than one-half as toxic as some western newts but are 10 times as toxic as the adult Eastern Newt.

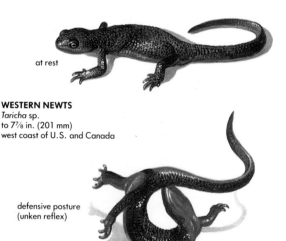

at rest

WESTERN NEWTS
Taricha sp.
to 7⅞ in. (201 mm)
west coast of U.S. and Canada

defensive posture
(unken reflex)

EASTERN NEWT
Notophthalmus viridescens
to 4⅜ in. (112 mm)
eastern third of U.S.

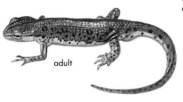

adult

terrestrial juvenile, eft stage

Blue Jay attacking eft

defensive posture

EUROPEAN NEWTS (22 species) all have skin toxins but are less toxic than western newts (p. 98). Most European and Asian newts have brightly colored bellies, which they expose when predators attack.

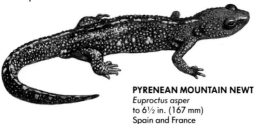

PYRENEAN MOUNTAIN NEWT
Euproctus asper
to 6½ in. (167 mm)
Spain and France

MARBLED NEWT
Triturus marmoratus
to 6½ in. (160 mm)
Spain and France

defensive posture

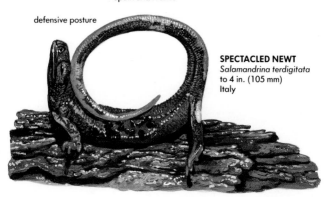

SPECTACLED NEWT
Salamandrina terdigitata
to 4 in. (105 mm)
Italy

ASIAN NEWTS (20 species) are probably all toxic at levels between European newts and western newts.

warts of skin glands

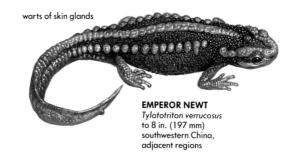

EMPEROR NEWT
Tylototriton verrucosus
to 8 in. (197 mm)
southwestern China,
adjacent regions

TAIL-SPOTTED NEWT
*Paramesotriton
caudopunctatus*
to 7 in. (170 mm)
China

SWORD-TAILED NEWT
Cynops ensicauda
to 6 in. (156 mm)
Japan

defensive posture

SPINEY NEWTS, found on the southern islands of Japan and the adjacent coast of China, have exceptionally long, branched ribs, their sharp tips piercing the skin through warts that consist of enlarged poison glands. The rib tips also carry small amounts of the poison when they jab into the mouth of an attacking predator. In humans, the venom causes a burning pain for at least 20 minutes if the ribs puncture the skin.

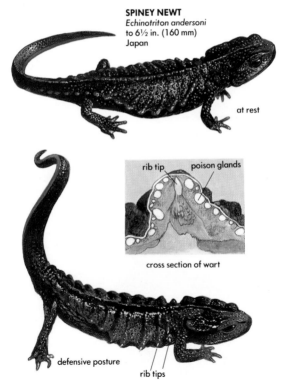

SPINEY NEWT
Echinotriton andersoni
to 6½ in. (160 mm)
Japan

at rest

rib tip   poison glands

cross section of wart

defensive posture

rib tips

FIRE SALAMANDERS spray toxic secretions (primarily samandarin) from greatly enlarged glands along the middle of its back. The salamander can direct the spray accurately to a distance of 6 feet. The spray contains a toxin that affects the central nervous system and burns the eyes and mouth. It is an effective protection from predators. Fire Salamander females migrate to streams to give birth to larvae rather than laying eggs like other salamanders and newts.

In mythology, these salamanders were believed immune to heat and fire. Their skin was believed to be made of asbestos, which cannot be burned. It was said also that people would die if they ate food over which these salamanders had crawled. Another belief was that touching the secretion of a Fire Salamander would cause a person's hair to fall out.

FIRE SALAMANDER
Salamandra salamandra
to 12½ in. (316 mm)
Europe to Israel and
North Africa

spray of neurotoxin

MOST SALAMANDERS produce skin secretions that are distasteful to would-be predators. If swallowed, the secretions burn the mouth and may cause nausea.

The salamanders present their most distasteful parts (usually the tail) to an attacker. This defensive behavior pattern and others are shown on these two pages.

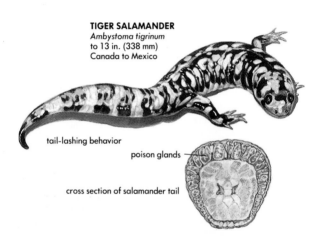

TIGER SALAMANDER
Ambystoma tigrinum
to 13 in. (338 mm)
Canada to Mexico

tail-lashing behavior

poison glands

cross section of salamander tail

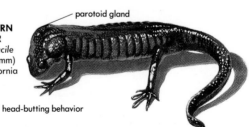

parotoid gland

NORTHWESTERN SALAMANDER
Ambystoma gracile
to 8½ in. (194 mm)
Alaska to California

head-butting behavior

tail-waving behavior

JEFFERSON'S SALAMANDER
Ambystoma jeffersonianum
to 8½ in. (196 mm)
eastern Canada to Kentucky

PACIFIC GIANT SALAMANDER
Dicamptodon ensatus
to 12 in. (300 mm)
British Columbia to California

body-arched behavior

tail-lashing behavior

ENSATINA SALAMANDER
Ensatina eschscholtzi
to 6 in. (150 mm)
British Columbia to California

tail-waving behavior

CAVE SALAMANDER
Eurycea lucifuga
to 7 in. (181 mm)
east-central U.S.

MEXICAN BOLITOGLOSSA
Bolitoglossa mexicanum
to 8 in. (192 mm)
Mexico to Honduras

coiling behavior

105

REPTILES

About 800 of the 6,500 species of reptiles in the world are venomous, their toxins produced by modified salivary glands. Venoms are injected by enlarged teeth, and their primary use is to subdue prey. None of the turtles, crocodilians, amphisbaenids, or the tuatara is venomous.

LIZARDS Only two species of venomous lizards occur in the world: the Gila Monster and the Beaded Lizard. Their venom glands are located along the outer edge of the lower jaw, and ducts empty the venom at the bases of greatly enlarged, grooved teeth. These lizards bite in self defense, holding on with a viselike grip and releasing venom into the wound. The venom causes excruciating pain, weakness, and dizziness but seldom death. The area around a bite remains tender for several weeks.

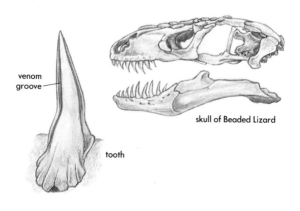

venom groove

tooth

skull of Beaded Lizard

GILA MONSTER
Heloderma suspectum
to 2 ft. (61 cm)
southwestern U.S.

BEADED LIZARD
Heloderma horridum
to 3 ft. 3 in. (100 cm)
western Mexico to Guatemala

SNAKES have been the subject of fascination, fear, and myths in all cultures. In the 1600s and before, the mythical cockatrice was considered the "king of snakes," its venom feared by all. It was believed to crawl with the front part of its body off the ground, and some believed it had wings.

Of the some 2,400 species of snakes in the world, at least 800 are venomous to some degree. An estimated 1,700,000 people are bitten by venomous snakes every year, and 40,000 to 50,000 of these bite victims die. In less industrialized countries many bites and deaths are not reported, and so this estimate of deaths is probably low. All the members of two families of snakes—Viperidae and Elapidae—are venomous. The family Colubridae contains both venomous and nonvenomous species. In other families, none of the members are venomous, the snakes having solid, ungrooved teeth (no fangs, or aglyphous).

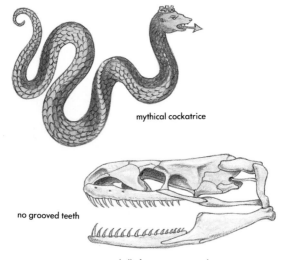

mythical cockatrice

no grooved teeth

skull of nonvenomous snake

COLUBRIDAE is a family of about 1,600 species, 400 of which are venomous to some extent. Because most of these snakes are small and their venom weak, they are not dangerous. The venom is produced by Duvernoy's gland, located behind the eye, and is transmitted through enlarged and grooved teeth at the rear of the mouth. For this reason they are known as rear-fanged (opisthoglyphous) snakes. Among the most dangerous of the venomous colubrids are the Boomslang and the Bird Snake, both of Africa, the Yamakagashi from the Orient, and the South American Hognose Snake, which has enormous rear fangs that can be erected.

Colubrids, worldwide distribution

skull of venomous colubrid

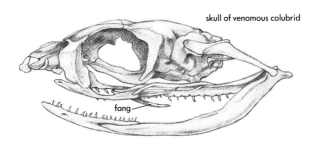

fang

ELAPIDAE is a family made up of front-fanged snakes, their deeply grooved or hollow fangs fixed in an erect position at the front of the mouth (proteroglyphous). Venoms produced by these snakes are primarily neurotoxic, but the venoms of some species also affect the heart. The bites of many of these snakes are no more painful than pin pricks, but the venom acts rapidly on the nervous system and causes death by suffocation when the respiratory system becomes paralyzed.

Coral snakes, cobras, kraits, mambas, and their Australian relatives are among the nearly 200 land-dwelling species in this family. Black Mambas are probably the most dangerous because of their size (nearly 14 feet), potent venom, and aggressive nature. Spitting Cobras have fangs with forward-facing openings from which they spray their venom into the eyes of victims.

The roughly 50 species of sea snakes are aquatic elapids. Some authorities classify them in a separate family: the Hydrophidae. The bodies and especially the tails are flattened side to side, enhancing their swimming ability. Sea snakes are most common in the South Pacific. Most species spend their entire lives in the water. Some are pelagic, forming rafts of snakes that may extend for miles in the open sea.

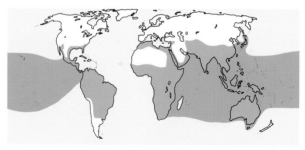

Elapids, worldwide distribution

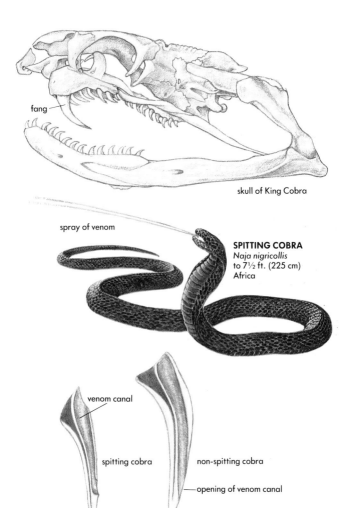

fang

skull of King Cobra

spray of venom

SPITTING COBRA
Naja nigricollis
to 7½ ft. (225 cm)
Africa

venom canal

spitting cobra

non-spitting cobra

opening of venom canal

111

VIPERIDAE is a family of stocky snakes with triangular heads. The rear swellings of the head house the venom glands, and the long, hollow fangs are located on the upper jaw bone, which can be rotated to erect the fangs when the snake strikes. When the mouth is closed, the fangs are folded back against its roof. This fang arrangement is called solenoglyphous. The venoms of vipers are primarily hemotoxic. Bites are painful and cause blistering, hemorrhaging, and digestion of tissue around the wound.

Vipers are often divided into two groups: true vipers (about 45 species) of Africa, Europe, and Asia, and the pit vipers, found primarily in the Americas but with a few species in Southeast Asia. Pit vipers are sometimes recognized as a separate family—the Crotalidae. Pit vipers, which include the rattlesnakes and moccasins, have a heat-sensitive pit between the nostril and the eye. These heat sensors allow the snakes to detect warm-blooded prey at considerable distances and also in the dark. Pit vipers direct their strikes by using the "image" formed by the heat sensors.

rattlesnake skull

fangs folded

Vipers, worldwide distribution

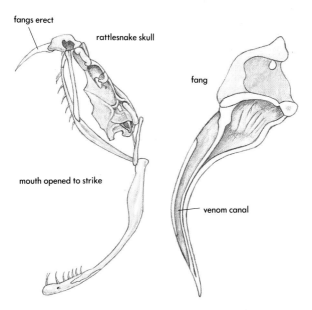

fangs erect

rattlesnake skull

fang

mouth opened to strike

venom canal

113

U.S. venomous snakes include 13 species of rattlesnakes, 2 pigmy rattlesnakes, 2 moccasins, 2 coral snakes, and at least 25 species of rear-fanged colubrids, of which only 2 are at all dangerous.

RATTLESNAKES (genus *Crotalus*) account for an estimated 7,000 bites per year in the United States, but only 9 or 10 of the bites are fatal. Most of the deaths are from bites of the Eastern Diamondback or the Western Diamondback. Western Rattlesnakes, Timber Rattlesnakes, and Sidewinders account for the largest number of bites, but few of the bites are fatal. The Mojave and Tiger rattlesnakes have a stronger venom than do most U.S. species, but they are responsible for only a few of the bites. Adjusted for number of people, the greatest number of bites occur in North Carolina, Arkansas, Texas, and Georgia.

EASTERN DIAMONDBACK RATTLESNAKE
Crotalus adamanteus
to 8 ft. (244 cm)
Gulf Coast and Atlantic seaboard from
eastern Louisiana to North Carolina

WESTERN DIAMONDBACK RATTLESNAKE
Crotalus atrox
to 7 ft. (213 cm)
California to Arkansas,
into Mexico

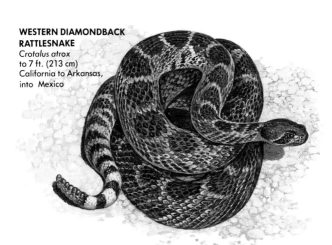

WESTERN RATTLESNAKE
Crotalus viridis
to 5 ft. 4 in. (163 cm)
Great Plains to Pacific Coast,
Canada to Mexico

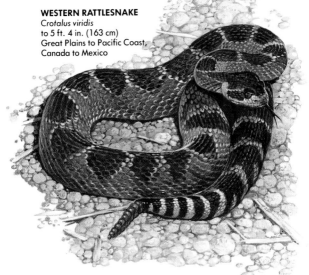

**TIMBER
RATTLESNAKE**
Crotalus horridus
to 6 ft. 2 in. (189 cm)
eastern Texas to Wisconsin,
New Hampshire, and Florida

SIDEWINDER
Crotalus cerastes
to 2 ft. 7 in. (79 cm)
deserts of southern California,
Arizona, and Nevada to Mexico

116

MOJAVE RATTLESNAKE
Crotalus scutulatus
to 4 ft. 3 in. (130 cm)
southern California,
Nevada, Arizona,
western Texas, into Mexico

TIGER RATTLESNAKE
Crotalus tigris
to 3 ft. (91 cm)
Arizona into Mexico

U.S. RATTLESNAKES

ROCK RATTLESNAKE
Crotalus lepidus
to 2 ft. 9 in. (83 cm)
southeastern Arizona to
western Texas, into Mexico

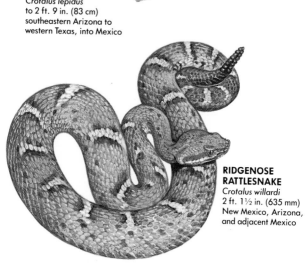

**RIDGENOSE
RATTLESNAKE**
Crotalus willardi
2 ft. 1½ in. (635 mm)
New Mexico, Arizona,
and adjacent Mexico

PIGMY RATTLESNAKES (genus *Sistrurus*) of two species occur in the United States. Their tails are more slender and their rattles narrower than rattlesnakes. These snakes are responsible for a number of painful bites every year, but fatalities are rare. A newly born pigmy rattlesnake has a bright yellow tail used to lure prey near enough to bite.

MASSASAUGA
Sistrurus catenatus
to 3 ft. 4 in. (100 cm)
New York and Michigan to
Texas and Arizona

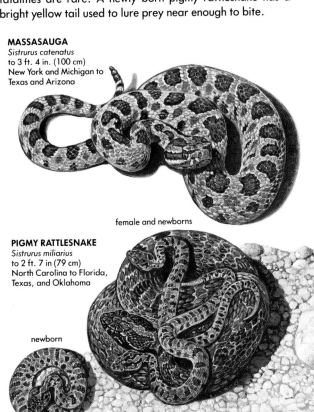

female and newborns

PIGMY RATTLESNAKE
Sistrurus miliarius
to 2 ft. 7 in (79 cm)
North Carolina to Florida,
Texas, and Oklahoma

newborn

MOCCASINS (genus *Agkistrodon*) are pit vipers with a venom similar to that of rattlesnakes, but moccasins do not have rattles. Worldwide, there are about a dozen species, most of them in Southeast Asia. One lives in Mexico. Two species—the Cottonmouth and the Copperhead—occur in the United States. The Cottonmouth causes an average of one death per year. The Copperhead is responsible for more bites than any other venomous snake in the United States, but deaths due to the bites are extremely rare.

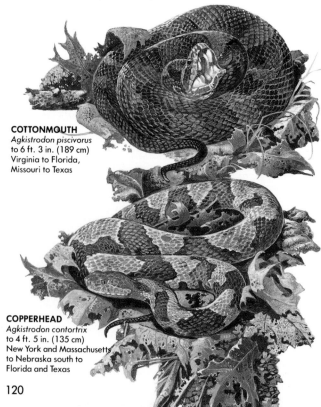

COTTONMOUTH
Agkistrodon piscivorus
to 6 ft. 3 in. (189 cm)
Virginia to Florida,
Missouri to Texas

COPPERHEAD
Agkistrodon contortrix
to 4 ft. 5 in. (135 cm)
New York and Massachusetts
to Nebraska south to
Florida and Texas

CORAL SNAKES of some 50 species belonging to the family Elapidae are restricted to the Americas. Only two species occur in the United States. Both have powerful neurotoxic venoms. The Arizona Coral Snake is generally nonaggressive, however, and no human deaths due to its bite are recorded. Several people, mostly in Texas and Florida, are bitten every year by Eastern Coral Snakes, and a death is recorded roughly every five years. The deaths are usually from respiratory paralysis and occur within 36 hours.

ARIZONA CORAL SNAKE
Microuroides euryxanthus
to 1 ft. 9 in. (53 cm)
Arizona and New Mexico into Mexico

Yellow bands are white in some individuals.

EASTERN CORAL SNAKE
Micrurus fulvius
to 4 ft (121 cm)
North Carolina and Florida
to Texas into Mexico

121

VENOMOUS REAR-FANGED SNAKES include more than 25 of the 92 species of colubrid snakes in the United States. They have grooved, enlarged teeth at the rear of the mouth, and they produce venoms used to paralyze their prey, often lizards. Most of these snakes have small mouths, and they are not dangerous to humans. The three species shown here—and especially large individuals—can give painful bites and should be handled with care. Many species in this family, such as racers, water snakes, kingsnakes, garter snakes, and bullsnakes, bite when captured but are not venomous.

NIGHT SNAKE
Hypsiglena torquata
to 2 ft. 2 in. (66 cm)
Washington to Nebraska,
into Mexico

CAT-EYED SNAKE
Leptodeira septentrionalis
to 3 ft. 3 in. (99 cm)
southern Texas,
into Mexico

LYRE SNAKE
Trimorphodon biscutatus
to 4 ft. (121 cm)
southern California to
Texas, into Mexico

EUROPE has only seven species of dangerously venomous snakes, and all are vipers. The most dangerous are the Nose-horned Viper, Ottoman Viper, and Blunt-nosed Viper. The last of these has even been reported to kill camels.

ADDER
Vipera berus
to 3 ft. (90 cm)
wide-ranging in Europe and
USSR, to Pacific Ocean

ASP VIPER
Vipera aspis
to 2½ ft. (75 cm)
primarily France and Italy

NOSE-HORNED VIPER
Vipera ammodytes
to 3 ft. (90 cm)
southeastern Europe and
southwestern Asia

OTTOMAN VIPER
Vipera xanthina
to 4 ft. (120 cm)
Turkey, also Asia Minor

BLUNT-NOSED VIPER
Vipera lebetina
to 5 ft. (150 cm)
Greek Islands, also
southwestern Asia,
northwestern Africa

123

INDIA is inhabited by 230 species of snakes, of which about 50 are venomous. The large number of venomous snakes and dense human populations result in an estimated 200,000 bites each year, from which an estimated 10,000 to 15,000 people die. Most of these bites are by only four species of snake: Russell's Viper, Saw-scaled Viper, Common Krait, and Common Cobra, the last probably accounting for the most deaths. The King Cobra, which has enough venom to kill an elephant and whose bite is normally fatal to humans, is uncommon, and bites to humans are rare.

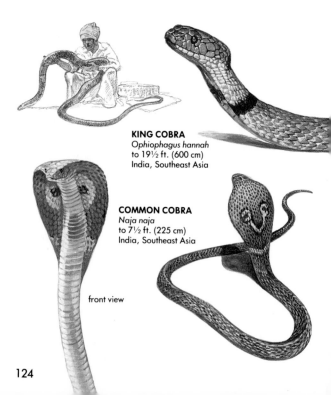

KING COBRA
Ophiophagus hannah
to 19½ ft. (600 cm)
India, Southeast Asia

COMMON COBRA
Naja naja
to 7½ ft. (225 cm)
India, Southeast Asia

front view

COMMON KRAIT
Bungaris caeruleus
to 5¾ ft. (173 cm)
India

BANDED KRAIT
Bungaris fasciatus
to 7 ft. (212 cm)
India, Southeast Asia

RUSSELL'S VIPER
Vipera russelli
to 5½ ft. (168 cm)
India, Southeast Asia

SOUTHEAST ASIA has many of the same venomous snakes as India, and snakebites are common. Burma apparently has the dubious honor of being the place in the world where one has the greatest chance of dying from snakebite. An estimated 15 per 100,000 people die yearly.

MALAYAN PIT VIPER
Calleselasma rhodostoma
to 3 ft. 4 in. (102 cm)
Southeast Asia

OKINAWA HABU
Trimeresurus flavoviridis
to 7½ ft. (225 cm)
Okinawa

SHARP-NOSED PIT VIPER
Deinagkistrodon acutus
to 5 ft. (150 cm)
China, Taiwan

FEA'S VIPER
Azemiops feae
to 3 ft. (90 cm)
mountains of
Southeast Asia

NORTH AFRICA has a rather limited number of venomous snakes (four elapids and ten vipers). These are species adapted to desert or grassland conditions. The most dangerous snakes in this area are probably the two species of Saw-scaled Vipers, one shown here and the other on p. 146.

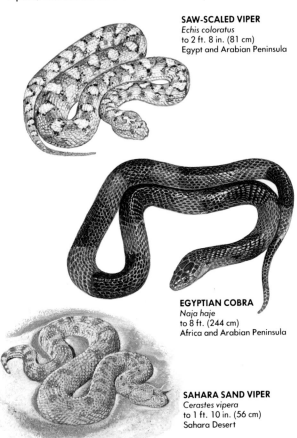

SAW-SCALED VIPER
Echis coloratus
to 2 ft. 8 in. (81 cm)
Egypt and Arabian Peninsula

EGYPTIAN COBRA
Naja haje
to 8 ft. (244 cm)
Africa and Arabian Peninsula

SAHARA SAND VIPER
Cerastes vipera
to 1 ft. 10 in. (56 cm)
Sahara Desert

AFRICA south of the Sahara Desert has about 300 snakes of the family Colubridae. Many of them are rear-fanged. Two of these, the Boomslang and the Bird Snake, are dangerously venomous. Both are arboreal snakes, and both can expand their necks in a threat display. In humans the bite causes extensive internal bleeding, sometimes resulting in death.

BOOMSLANG
Dispholidus typus
to 6 ft. (183 cm)
Africa (south of Sahara)

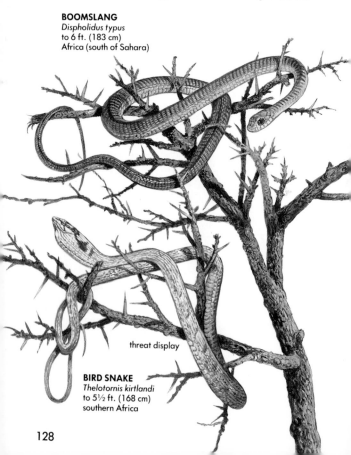

threat display

BIRD SNAKE
Thelotornis kirtlandi
to 5½ ft. (168 cm)
southern Africa

128

ELAPID snakes of about 20 species also inhabit Africa south of the Sahara. Most of these have dangerously neurotoxic venom, and several are "spitters" that are able to spray their venom into the eyes of people. The venom can cause blindness if not removed quickly. The Black Mamba is the longest and most feared of Africa's venomous snakes.

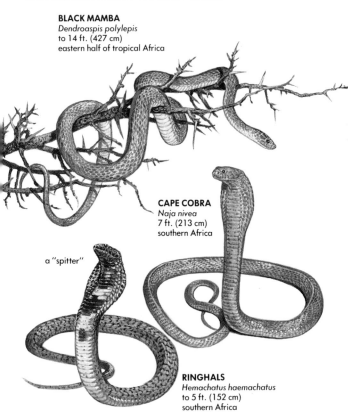

BLACK MAMBA
Dendroaspis polylepis
to 14 ft. (427 cm)
eastern half of tropical Africa

CAPE COBRA
Naja nivea
7 ft. (213 cm)
southern Africa

a "spitter"

RINGHALS
Hemachatus haemachatus
to 5 ft. (152 cm)
southern Africa

VIPERS of about 30 species inhabit Africa south of the Sahara Desert. Many of these vipers are small or uncommon, but the large vipers are very dangerous. The Puff Adder probably kills more people than any other African snake. The Gaboon Viper may have a head width of 5 inches, with fangs 2 inches long.

PUFF ADDER
Bitis arietans
to 5 ft. (152 cm)
all of Africa except
forests and deserts

GABOON VIPER
Bitis gabonica
to 6 ft. 8 in. (204 cm)
forest areas
south of Sahara

AFRICAN BUSH VIPER
Atheris squamiger
to 2 ft. 5 in. (78 cm)
forests of tropical Africa

RHINOCEROS VIPER
Bitis nasicornis
to 4 ft. (122 cm)
rain forests of
central Africa

131

AUSTRALIA has some of the world's most dangerous elapids (about 75 species) such as the Death Adder, Taipan, Tiger Snake, and King Brown Snake. Because of excellent anti-venin production in Australia, however, there are only about five human deaths per year.

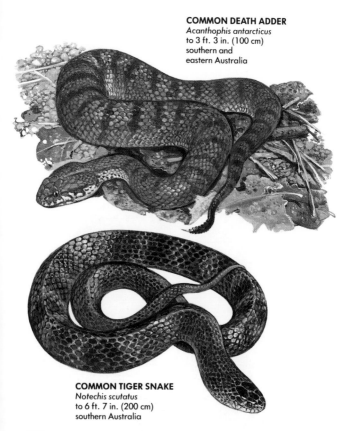

COMMON DEATH ADDER
Acanthophis antarcticus
to 3 ft. 3 in. (100 cm)
southern and
eastern Australia

COMMON TIGER SNAKE
Notechis scutatus
to 6 ft. 7 in. (200 cm)
southern Australia

TAIPAN
Oxyuranus scutellatus
to 6 ft 7 in. (200 cm)
northern and
eastern Australia

KING BROWN SNAKE
Pseudechis australis
to 8 ft. 10 in. (270 cm)
Australia

RED-BELLIED BLACK SNAKE
Pseudechis porphyriacus
to 8 ft. 2 in. (250 cm)
eastern Australia

SEA SNAKES are most common in the coastal waters of Australia and southern Asia, with one species ranging across the Pacific Ocean to Central and South America. Most of these are highly venomous but mild tempered.

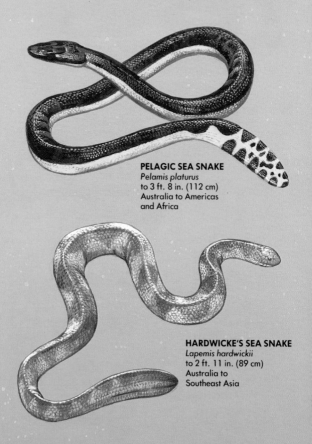

PELAGIC SEA SNAKE
Pelamis platurus
to 3 ft. 8 in. (112 cm)
Australia to Americas
and Africa

HARDWICKE'S SEA SNAKE
Lapemis hardwickii
to 2 ft. 11 in. (89 cm)
Australia to
Southeast Asia

BANDED SEA SNAKE
Laticauda colubrina
to 4 ft. 7 in. (140 cm)
Australia to
Southeast Asia

BLACK-HEADED SEA SNAKE
Hydrophis melanocephalus
to 3 ft. 3 in. (100 cm)
Australia

SEA SNAKE
Aipysurus apraefrontalis
to 1 ft. 8 in. (50 cm)
Australia

MEXICO has more species of venomous snakes than any other country in the Americas. The elapids include 1 sea snake and 14 coral snakes. The vipers are represented by 24 rattlesnakes, 2 pigmy rattlesnakes, 2 moccasins, and 16 other pit vipers. In addition, there is an undetermined number of venomous rear-fanged colubrids.

BLACKTAIL RATTLESNAKE
Crotalus molossus
to 4 ft. 2 in. (126 cm)
Arizona, New Mexico, and
Texas, into Mexico

**MEXICAN WEST-COAST
RATTLESNAKE**
Crotalus basiliscus
to 6 ft. 7 in. (200 cm)
western Mexico

MEXICAN SMALL-HEADED RATTLESNAKE
Crotalus intermedius
to 2 ft. (60 cm)
central Mexico

MEXICAN PIGMY RATTLESNAKE
Sistrurus ravus
to 2 ft. 4 in. (70 cm)
central Mexico

VARIABLE CORAL SNAKE
Micrurus diastema
to 2 ft. 10 in. (85 cm)
central Mexico to Honduras

CENTRAL AMERICA'S venomous snakes total 33 species: 16 elapids and 17 pit vipers (1 rattlesnake, 1 moccasin, and 15 others). The Barba Amarilla is responsible for the most bites, many of them fatal.

NEOTROPICAL RATTLESNAKE
Crotalus durissus
to 6 ft. (180 cm)
Mexico into South America

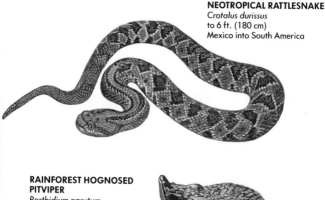

RAINFOREST HOGNOSED PITVIPER
Porthidium nasutum
to 2 ft. (60 cm)
southern Mexico to Colombia

CANTIL
Agkistrodon bilineatus
to 4 ft. 7 in. (138 cm)
Pacific Coast of Mexico
to Costa Rica

GUATEMALAN PALM PITVIPER
Bothriechis bicolor
to 2 ft. 4 in. (70 cm)
southern Mexico to
Honduras

BARBA AMARILLA
Bothrops asper
to 8 ft. 2 in. (250 cm)
Mexico into South America

JUMPING PITVIPER
Porthidium nummifer
to 2 ft. 7 in. (80 cm)
southern Mexico to Panama

139

YELLOW-BLOTCHED PALM PITVIPER
Bothriechis aurifer
to 3 ft. 4 in. (101 cm)
southern Mexico and Guatemala

EYELASH PALM PITVIPER
Bothriechis schlegeli
to 2 ft. 7 in. (80 cm)
southern Mexico into
South America

yellow phase

dark phase
threat display

CENTRAL AMERICAN CORAL SNAKE
Micrurus nigrocinctus
to 3 ft. 3 in. (100 cm)
southern Mexico to Panama

ALLEN'S CORAL SNAKE
Micrurus alleni
to 3 ft. 6 in. (107 cm)
Nicaragua to Panama

MANY-BANDED CORAL SNAKE
Micrurus multifasciatus
to 4 ft. (120 cm)
Nicaragua to Panama

141

SOUTH AMERICA has 83 species of dangerously venomous snakes. Elapids are represented by 37 species of coral snakes and 1 sea snake; pit vipers by 45 species, two of which are rattlesnakes. An unknown number of colubrids from this region, and the rest of the world, are venomous, but most are not dangerous.

TWO-STRIPED FOREST PITVIPER
Bothriopsis bilineata
to 3 ft. 3 in. (100 cm)
South America

BUSHMASTER
Lachesis muta
to 12 ft. (360 cm)
Central and
South America

COMMON LANCEHEAD
Bothrops atrox
to 6 ft. 7 in. (200 cm)
South America

SOUTHERN CORAL SNAKE
Micrurus frontalis
to 4 ft. 5 in. (135 cm)
South America

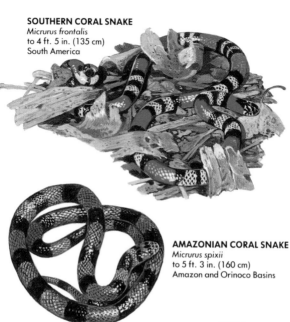

AMAZONIAN CORAL SNAKE
Micrurus spixii
to 5 ft. 3 in. (160 cm)
Amazon and Orinoco Basins

tail in defensive display

AQUATIC CORAL SNAKE
Micrurus surinamensis
to 5 ft. 11 in. (180 cm)
South America

CORAL SNAKE MIMICRY occurs in many species of colubrid snakes in North, Central, and South America. These are nonvenomous or rear-fanged species, and the coral snake resemblance, known as mimicry, is an advantage to the mimics because predators avoid the brightly banded coral snakes. Migratory birds may learn to avoid coral snakes in Central or South America and then avoid certain king snakes in regions of the United States where there are no coral snakes.

MILK SNAKE
Lampropeltis triangulum
to 4 ft. 4 in. (132 cm)
eastern two thirds
of U.S. to Central
America

Populations in the same areas as coral snakes have banded patterns; most other populations do not.

nonvenomous

nonvenomous

CALIFORNIA MOUNTAIN KINGSNAKE
Lampropeltis zonata
to 3 ft. 4 in. (102 cm)
California and southwestern Oregon

SCARLET SNAKE
Cemophora coccinea
to 2 ft. 8½ in. (83 cm)
southeastern U.S.

nonvenomous

CORAL SNAKE MIMICS FROM CENTRAL AND SOUTH AMERICA

Scaphiodontophis annulatus
to 2 ft. 7 in. (78 cm)
Honduras

Sibon sartorii
to 2 ft. (60 cm)
Guatemala

Lystrophis semicinctus
to about 2 ft. (60 cm)
southern South America

Tantilla annulata
to about 1 ft. 4 in. (40 cm)
Costa Rica

VIPER MIMICRY also occurs throughout the world. These are harmless species of snakes similar enough in color pattern and behavior to be mistaken by people and probably predators for various vipers. In Africa, Egg-eating Snakes even mimic the behavior and sound of Saw-scale Vipers, which make a rasping sound by rubbing their scales together.

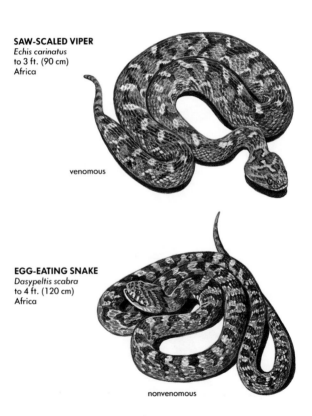

SAW-SCALED VIPER
Echis carinatus
to 3 ft. (90 cm)
Africa

venomous

EGG-EATING SNAKE
Dasypeltis scabra
to 4 ft. (120 cm)
Africa

nonvenomous

RATTLESNAKES are mimicked by snakes such as gopher snakes and hognose snakes. These harmless snakes vibrate their tails, which in dry grass or leaves results in a sound that resembles the rattle of rattlesnakes. Burrowing Owls hiss from their burrows with a sound like a rattlesnake rattle— frightening off animals like foxes and badgers.

The hiss of a Burrowing Owl sounds a lot like the frightening rattle of a venomous rattlesnake.

venomous

WESTERN RATTLESNAKE
Crotalns viridis
to 5 ft. 4 in. (163 cm)
Great Plains to Pacific
Coast, Canada to Mexico

WESTERN HOGNOSE SNAKE
Heterodon nasicus
to 3 ft. (90 cm)
Great Plains states

nonvenomous

MAMMALS

Only a few mammals are venomous. Monotremes have venomous spurs, and some shrews and their relatives have venomous saliva. Tenrecs and hedgehogs annoint their spines with toxins of other animals.

MONOTREMES are primitive mammals that lay eggs. All living monotremes (the Platypus and the echidnas) have hollow spurs on the inner side of their lower hind legs. In the male Platypus, the half-inch-long spurs can be erected, and the venom they release into a wound causes pain and swelling for several days. Dogs have been killed by this venom, but no human deaths have been reported. The spurs may be used for territorial fights with other males, to subdue large prey, or to repel predators. Echidnas have a similar spur, but their venoms have not been studied.

ECHIDNA
Tachyglossus aculeatus
to 18 in. (45 cm)
weight to 17½ lbs. (8 kg)
New Guinea and Australia

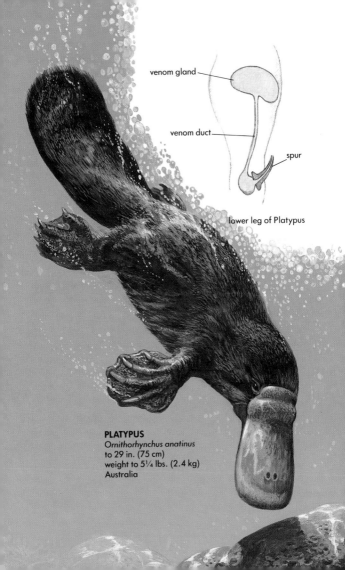

venom gland

venom duct

spur

lower leg of Platypus

PLATYPUS
Ornithorhynchus anatinus
to 29 in. (75 cm)
weight to 5¼ lbs. (2.4 kg)
Australia

SHREWS are fierce predators that use their venomous saliva to immobilize mice and other animals larger than themselves. Shrews then store their prey, sometimes alive but paralyzed, to be eaten later. The venom is transmitted into a bite wound along a groove formed by the lower incisor (front) teeth. The venom has different components. One affects cold-blooded prey, such as insects; another mice, or other warm-blooded animals. Shrew bites are not dangerous to humans but can cause a burning pain that lasts for hours. The Short-tailed Shrew of North America is about three times more venomous than the European Water Shrew.

SHORT-TAILED SHREW
Blarina brevicauda
to 6 in. (160 mm)
eastern U.S. and adjacent Canada

food cache of frogs

EUROPEAN WATER SHREW
Neomys fodiens
to 7 in. (75 mm)
Europe

These shrews capture prey both
underwater and on land.

SHREWS of about 250 species occur everywhere in the world except in Australia and southern South America. Some are very small, weighing less than an ounce and only about an inch and a half long. One species measures more than 8 inches long. The toxicity of the saliva of most species has not been tested.

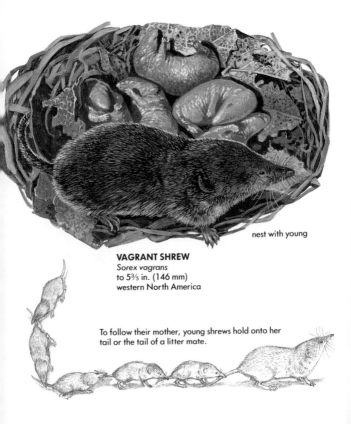

nest with young

VAGRANT SHREW
Sorex vagrans
to 5⅗ in. (146 mm)
western North America

To follow their mother, young shrews hold onto her tail or the tail of a litter mate.

SOLENODONS are larger relatives of shrews and are also mildly toxic. The two species occur only in Haiti, the Dominican Republic, and Cuba. Both are endangered. Their venom, which is much less toxic than that of a Short-tailed Shrew, is conducted up grooves on the rear surface of their elongated lower incisor teeth. Because of their relatively large size and the greater volume of venom, their bites are painful.

lower incisors

HAITIAN SOLENODON
Solenodon paradoxus
to 23 in. (58 cm)
Haiti and Dominican Republic

HEDGEHOGS (12 species) live in Europe, Africa, and Asia. They protect themselves by anointing their spines with the poisonous secretions of other animals. Before eating a toad, a hedgehog chews the secretions from the toad's skin into a froth that is then licked onto the spines. When attacked, hedgehogs curl into a ball, protecting the head and belly. They also hiss and jump, jabbing the spines coated with dried toxins into the attacking predator. This greatly increases the pain of being jabbed by the spines and the chance of infection. They represent no great danger to humans. Some tenrecs from Madagascar exhibit the same behavior as hedgehogs.

rubbing toad on spines

LONG-EARED DESERT HEDGEHOG
Hemiechinus auritus
to 12½ in. (320 mm)
India, Pakistan, Egypt

AFRICAN HEDGEHOG
Erinaceus albiventris
to 13½ in (350 mm)
central Africa

licking poison and saliva onto the spines

Young hedgehogs lick toxins from their mother's spines and
self-anoint themselves even before their eyes are open.

BOOKS FOR FURTHER STUDY

Campbell, Jonathan A., and William W. Lamar. *Venomous Reptiles of Latin America.* Cornell University Press, Ithaca, NY, 1989.

Caras, Roger. *Venomous Animals of the World.* Prentice Hall, Englewood Cliffs, NJ, 1974.

Freiberg, Marcos A., and Jerry G. Walls. *The World of Venomous Animals.* TFH Publications Inc., Neptune City, NJ, 1984.

Habermehl, Gerhard G. *Venomous Animals and Their Toxins.* Springer-Verlag, New York, 1981.

Halstead, Bruce W. *Poisonous and Venomous Marine Animals of the World.* 2nd ed. rev. Darwin Pr., Inc., Princeton, NJ, 1978.

Minton, Sherman A. Jr., and Madge R. Minton. *Venomous Reptiles.* rev. ed. Charles Scribner's Sons, New York, 1980.

Prins, Andre, and Vincent Leroux. *South African Spiders and Scorpions.* Anubis Press, Cape Town, 1986.

Smith, Hobart M., and Edmund D. Brodie, Jr. *Reptiles of North America,* A Golden Field Guide. Golden Press, New York, 1982.

Sutherland, Struan K. *Venomous Creatures of Australia.* 2nd ed. Oxford University Press, New York, 1985.

MAGAZINES with occasional popular articles on venomous animals include: *National Geographic, Natural History, National Wildlife,* and *International Wildlife.*

ZOOS always have collections of live venomous snakes, and public aquaria often display venomous fishes. Few zoos display other venomous animals—outstanding exceptions that maintain live collections of venomous invertebrates are the National Zoological Park in Washington, D.C. and the Arizona-Sonora Desert Museum in Tuscon, Arizona.

INDEX

MEASURING SCALE (IN MILLIMETERS AND CENTIMETERS)